The Science of
FULL MOON INVOCATIONS
from Humanity's Heart to Hierarchy's Will

THE DIVINE CONCORDANCE OF LIGHT III

VOLUME 1

Dadi Darshan Dharma

Printed and bound in Canada by Imprimeries Transcontinental, April 2007

The Science of Full Moon Invocations, from Humanity's Heart to Hierarchy's Will — The Divine Concordance of Light III. Volume 1. Dadi Darshan Dharma.

ISBN: 978-0-9734439-0-5

© 2007 Magnificent Magus Publications

Registration of copyright: Second trimester 2007
National Library of Quebec
National Library of Canada

Magnificent Magus Publications©
235 Rene Levesque Boulevard East, Suite 310, Montreal, Quebec, Canada H2X 1N8
Telephone: (514) 255-8700 ~ Facsimile: (514) 255-0478
E-mail: info@palmpublications.com
Web site: http://www.palmpublications.com

Graphic design: D.D.D., Eric Mathieu, Lucie Robitaille
Typesetting: Louise Roy
Photographs by Simhananda

© All rights reserved. No part of this book may be reproduced in any form without permission in writing from the author, except to quote or photocopy specific passages for the purposes of group study.

Publications by Magnificent Magus Publications:

Seven Studies of Soul Stations or Soul-ar Progressions Upon Each of the Seven Cosmic-Physical Rays (an integral excerpt from Collectanea One, *The Divine Concordance of Light*). Etbonan Karta. 2007.

Seven Sacred Stations of the Self & Seven Flaming Fiats of Light Upon the Seven Cosmic-Physical Rays (an integral excerpt from *The Divine Concordance of Light*). Etbonan Karta. 2001.

The Divine Concordance of Light: A Handbook from Heaven to Progression Earth — "The Seven Rays of God: Seven Studies of the Soul's Earthly Pilgrimage of Service Upon the Seven Cosmic-Physical Rays". Etbonan Karta. 2001.

Forthcoming books:

Scriptings of the Soul in Questions of Light. Dadi Darshan Dharma.

The Divine Concordance of Light II: The Science of Invocation and the Art of Affirmation from Station HUMANITY to HIERARCHY's Heart. Etbonan Karta.

Publications by Orange Palm Publications:

Buddhas, Bodhisattvas, Khadromas and the Way of the Pilgrim — A Transformative Book of Photography and Pithy Sayings. Simhananda. 2007.

Holy-Moly Hiccoughs and Enigmatic Knotty Eructations From the Boffola Belly of Bu'Tai. The Drôleries and Dictums of Crazy Modern Dzog-zen. Ken N.O. Sho. 2007.

Knots of Eternity — Paradoxes from Dadi to Daughter. Volume 1. Dadi Darshan Dharma. 2007.

The Smiling Forehead — Paradoxes from Dadi to Daughter. Volume 2. Dadi Darshan Dharma. 2007.

The Great Golden Garland of Gampopa's Sublime Considerations on the Supreme Path — Contemplative Contemporary Commentaries of Gampopa's Root Text. Volume 1. B. Simhananda. 2005.

Paradisal Plums: Peaceful Ponderings from a (Rebel) Pandit's Puce Palm — Aphorisms, Adages, and Analects of Sri Adi Dadi, Volumes 1, 2. Etbonan Karta. 2001.

Forthcoming books:

Flyers from the Boys in the Buddhafield. B. Simhananda.

Paradisal Plums: Peaceful Ponderings from a (Rebel) Pandit's Puce Palm, — Aphorisms, Adages, and Analects of Sri Adi Dadi, Volumes 3, 4. Etbonan Karta.

Table of Contents

	The Art and Science of Invocation (and of Evocation)	
January:	Morya's Moon	3
February:	Upon the Anvil of the Primordial Lord's Exalted Will A Full Moon Invocation upon Ray I	7
March:	Immortal Thanks to Hierarchy	11
	March Full Moon	13
	Part I: O Buddha Fire and the Flame of Christ. Part II: O Maitreyic Christ	
April:	April or Easter, Full Moon. Part I: The Risen, Living Christ	17
	Part II: The Loving, Living Christ	18
	Part III: Litany to Buddha Maitreya, the Christ Lord	19
May:	Buddha Full Moon (The Month of May Invocation)	23
	Section II: 131 Names of Buddha – Wesak	24
	Section III: The Buddha-to-Christ Transmission (Upon Ray I)	30
	Section IV: Transmission of the Will-To-Good upon the Sub-Rays (of Ray I) to Humanity	31
June:	He Comes (The June Full-Moon Invocation)	35
	Sectioin Two: Festival of the June Full Moon	41
July:	Full Moon of July — "Lotus of Light"	49
	July Full Moon — Alternate Invocation	51
August:	Magical Moon of Krishna's Flute	55
	Section II: August Full Moon (Alternate Invocation)	57
September:	Ganesh Chaturthi: Part I Introduction, Geographical Traveller	61
	Part II: Historic Glory of Ganesha	62
	Part III: Invocative Celebration of Nine Ganapati Mantric Pilgrimages Leading to GANESHA CHATURTHI	63
	Part IV: 'OM Srinathadi Guru Ganapatim Namah' 'OM, salutations to the Primordial Guru, to Ganapati, (I bow)'	67
October:	Devi Durga Puja — October Autumnal Full Moon	73
	Part 1: Birth of the Devi	74
	Part 2: The Devi at War with the Demons	81
	Part 3: Invocatory Conjurations to Mahadevi Durga	89
November:	Kwan Shih Yin — November Autumnal Full Moon, Section I	107
	Part 1: Kwan Yin, Mother of Mercy	108
	Part II: Kwan Yin's Liberation	111
	Part III: Kwan Yin's Instantaneous Response to Invocation and the Immediate Benefits	113
	Part IV: Kwan Yin's Luminous Forms, Variegated Vahanas and Noble Ornaments	116
	Part V: Kwan Shih Yin Sweet Savioress of Mankind	123
	Section II : Invocation to Mary Motherly Matrix of the Divine	126
December:	I. December Full Moon — A Christmas Celebration	131
	II. In the December Full Moon of the Festive Season	131
	In the December Full Moon of the Festive Season (for Buddhists)	133

The Art and Science of Invocation (and of Evocation)

Everything that exists within our solar system is governed and qualified by different energies referred to as Physical Cosmic Rays. There are seven of them and they are to be differentiated as follows:

Ray 1, Will or Power
Ray 2, Love-Wisdom
Ray 3, Active Intelligence
Ray 4, Harmony through Conflict
Ray 5, Science or Concrete Knowledge
Ray 6, Devotion or Idealism
Ray 7, Order or Ceremonial Magic

For each New Age, a new expansion of energy is manifested and Humanity is greatly governed by it. New changes in attitude and fresh ways of approaching situations necessarily come into being and necessarily qualify everything. During the past Piscean era, the predominant energy was qualified by the Ray 6. In this past era, prayer was the formal method used in requests for help, be it spiritual, emotional or physical. People prayed for their own healing, or for the healing of loved ones. But, in the present Aquarian Age, with the dynamic activation of Ray 7, everyone now needs to re-learn how to pray, or rather, intelligently learn how to rightly Invoke. The penchant for individualized prayer, must now make room for a group-centered approach, which must necessarily be more mentally-polarized.

The needs of Humanity are changing and the way of doing things must adapt itself to these evolutionary changes. For the ongoing welfare of humanity and for its progressive evolution, a group-approach and a mentally-polarized focus must be made, in order to request the requisite help and support needed from the Hierarchy of Masters.

The Holy Hierarchy consists of a group of Realized Beings who have already tread the Path prior to us and have attained to Complete Mastery on the inferior dimensions, or lesser planes of Livingness. The art and science of Invocation represents the radically new, albeit ancient, mode of approach in summoning the Masters for assistance, guidance, direction, or Divine Intercession.

The Full Moon Invocations that are to be recited during the special night of the appropriate full moon, serve to establish an authentic soul-ar group contact, and develop a closer relationship between the Centre of Humanity and that of the Holy Hierarchy. Invocations have the amplificatory effect of raising the group consciousness, and of humanly promoting a more precise alignment with Hierarchy.

Full Moon Invocations are in effect, a Call to the higher energies, as well as being a superior scaffolding for the expression of the Verb, and are used for the primary purpose of stimulating in Holy Hierarchy an *evocative* response, so that Spiritual Light, Love, Liberty, Beauty and Good Will can be made to outpour unto Mankind, and subsequently, made to resonate within each man and woman's head and heart.

The evocation of Hierarchy's highly spiritual, energetic currents can positively stimulate the human consciousness, by imprinting on the mental continuum, certain fundamental ideas and essential ideals which are considered as necessary, for any substantial progress to be realized. In this way, a new medium of wholesome living can eventually be established, one that essentially promotes a right kind of relationship between nations and peoples, as well as in the various realms, or kingdoms, of Nature.

The Art of Invocation is a science. It is based fundamentally on the power of the Word and the power of intentioned, or directed thought.

In order for group work to be effective, thinking must be basically clear and vitally free of all negativity. It has to be detached, concentrated, and focused.

Adequate preparation and disciplined training are, therefore, considered necessitous, so that each invocative group becomes a pure receptacle with a backbone strong enough and ready enough to receive the invoked energies… which are to be transmitted by Hierarchy during the organized Full Moon meditation.

Ideally, the group should always receive the energies which are correctly 'invoked'. In knowing this, and applying the invocation in a scientific or disciplined manner, group work ought to become one of the greatest forces, or tools of liberation, for the whole of future Humanity.

"The Spiritual Science of Group Invocation,
And the Occult Practice of Sacred speech,
Will become the *sine qua non* tool
Of the evolved Consciousness
In the Aquarian Age."

D.D.D.

January Full Moon

El Morya Khan

Master El Morya Khan is the Chohan Lord of Ray 1. He transmits the energy of Power and of Divine Will. He works in close collaboration with the different Masters of the Great White Brotherhood, among whom is the Lord Kuthumi and the Adept, Lord St. Germain.

One of his primary tasks is to inspire men of State all over the world, so that they can make the necessary decisions and take the appropriate actions that will veer human evolution toward a superior foundation of better relationships and a greater harmony between the Heart of Humankind and the Heart of Hierarchy.

The Spire of Divine Will Fused to the Cathedral of the Heart (The Angel of Gethsemane)

Rays 1
3
2

El Morya Khan

Morya's Moon

El Morya Khan, our sangha, (or united group), calls proudly upon Thee now
In animated joy and conscious celebration of this new year's first Full Moon.

We impetrate Thee to teach us the occult ways of the adamantine Blue WILL OF GOD.

We impassionately importune Thee to catapult us fast upon wings of Light,
Straight, direct, and true as an arrow to the target of Thy Temple's Inner
Chambered BRIGHT.

We plead, ply and implore Thee, to fling our Soul's Intent implacably
Upon the sacrificial altar of the Ray I CHOHAN — Lord MORYA's flaming breast.

Brand us, O merciful Morya, with the Mahayana Seal of COMPASSION's perfect
Fire Crest.

Garner from this present Invocatory Group the total of us ablaze with the best of us.

Fill us perfectly flawless in Will, and infuse us with the fiery fervor of SANAT KUMARA's
Breath.

Extirpate from reincarnation's cyclical samsara the encrusted cornucopia of our
manifold cacophonous deaths.

Wrest from us, perdurably, the bittersweet incantations of Mara's mayic tresses,
and of Time's multifold, karmic tests.

Draw out, and ignite from the core of us the cleanest Call of GOOD for all
beings sentient and godly.

And deeply plant into the soil of our United Group, the Power-Seed of WISDOM's
Highest Truth enthralled.

Pitilessly pierce our Group Heart through, and through, and through, with the
Imprint Divine of Thy Seal Sublime.

Finely wrap around us the Manu Mantle of Thy stately, worldly service and
Thy great Bodhisattva caring.

May this Full-Moon Invocation, clear cut into our keenly astute Group-Consciousness,
the unique Ecstasis of Eternity.

And may we now ride bareback and wild upon the white-hot, whirling, wailing
winds of Light,

As they flow swiftly and tall from the dark, radiant beauty of Thy GOD-filled eyes,
and bright, Empty Breast.

February Full Moon

Sanat Kumara, the Primordial Lord

He has been called the Ancient of Days, the Youth of Endless Summers, the Lord of the World, the One Initiator. Sanat Kumara is the LORD of Shamballa and presides over the Great White Lodge of Hierarchical Masters. He is the unique manifestation and representative of the Planetary Logos.

The Heart of Love
and the Primordial Lord's Will

Rays 2
1
3

Sanat Kumara
The Primordial Lord

Upon the Anvil of the Primordial Lord's Exalted Will
A Full Moon Invocation upon Ray I

From the potent, silvery orb of the Primordial Lord's shimmering full moon
And via the filtering auspices of friendly February's familiar heartfelt reflection,

Let the exalted Point of Power which lies behind the Central Sun
Shine forth the Light Electric of the dark-hooded, Adamantine One.

Let His blue lightning streak dagger-like through space and shred the sky asunder.
Let His red fire crackle tonitruantly across the earth like howling rippling thunder.

Let the crepitant encircling inferno of His living, swarthy Flame
Usher everything crashing in through the Door of Death in game.

Let this group be willing to be destroyed upon the Anvil of That WILL
And may we be re-hammered and Pure Molded upon that Holy Hill,

Where there, occultly hidden behind the million Glories of the Brilliant Sun,
Lonely sits the GLORIOUS ONE, Who cryptically incarnates the Sacred Seven upon the Throne.

Where there, in consummate secrecy the Keys of Heaven are arcanely held,
Unfolding forth the Circle of Life even unto the obscured encirclement of Hell.

Let the 'One about Whom Naught Can Be Said', nor anything ever be known,
Save maybe that of distant echoed rumblings of Absolute Power yet unwound.

Let He, Whose Primal Word of Fohatic Fire spews forth the Infinite Ocean of Creation,
Elicit a legion of responsive, faithful Fiats of O.M. from the parched throat of renunciate man,
Thereby killing Desire, and soliciting forth a fiery downward look at Mankind, from the esteemed brow of the Supreme LORD, SANAT KUMARA.

Let He, Whose singeing Single Ray purely burns to impotent dross all imperfection,
And Whose Solitary Sound shatters irrevocably the Form, and compassionately cleaves the Absolute Emptiness,

Uplift us, one by one, quintessentially inspired into the magnificent Burning of the BRIGHT.

May the Primordial LORD stamp unto our Forehead the sacred searing of the SOLITARY ONE's High Purpose, which saves the Life.

And may He Who Is the WILL that Rules, give everyone the rising of the Gift of LOVE, in the dying embers of this great-hearted, February Moon's, exalted night.

March Full Moon

The Christ, Head of Hierarchy

The Maitreyic Christ is the Living Head of the Holy Hierarchy, the Master of Masters and of Angels alike, and the Chief Instructor of the World. The Energy of Love that pours from the Heart of the High Lord, (or God), is transmitted to Humanity through Him. He incarnates into the universe, the Christic Logoic principle. To Him especially, has been confided the task of directing the spiritual destiny of Mankind during the great Golden Age of Aquarius, through the stimulation of all hearts within Humanity and the spiritual upliftment of their consciousness.

Holy Hierarchy, Buddha Fire and the Flame of Christ

Rays
3
2
1

The Christ
Head of Hierarchy

Immortal Thanks to Hierarchy

He is amongst us at this time,
And will be for the next two thousand years,
Serving in the capacity of Chief Adept
And Acting Head of the *Physical Cosmic Ashram*,
Known to us as the 'Hierarchy of the Great White Brotherhood'.

Presently, through the Lighted-Focus
Of this March, Full Moon Dedication,
We do hereby take the opportunity to give
Thanks to the Christ and to His Holy Hierarchy,
For both the Fact and Truth
Of their Divine Existence upon this pilgrim planet.

We honor and give thanks
To Holy Hierarchy's past and present Devotion
Toward all of the earthly kingdoms;
And especially, for Its ongoing Dedication and Service
Toward Humankind's progressive evolution,
Both in the physical phenomenal sense,
As well as in the Spiritual mode of expression.

Let us, therefore, show appreciation for the Hierarchical Masters,
Who selflessly hold Humanity's hand,
As though they were gently guiding a Cosmic Child;
And let us honor and pay heed
To their gently guiding us with Light, Love and Life,
From Their Alpha-point in space to our Omega-terminal in time.

Let there be a profound gratitude proffered to Them,
From the Chalice of our awakening Consciousness.

Let there be heaps of Immortal Thanks offered to Them,
From the humble Tabernacle of our Devotional Heart.

Let us rest assured and be made aware,

March Full Moon

That the Masters of Wisdom will continue
In an ever-increasing precipitation of Light,
To help Mankind, in its long-suffering trek
Toward the fulfillment of our Planet's "Manifest Destiny".

Purely dedicated and sacrificially forsworn,
Have the Hierarchy of Masters always been,
Towards helping the evolving Earth, to progressively
Become a 'Sacred Planet' within the magnificent *Great Plan*,
And the even *Greater Purpose of the Logoic S*ANAT *K*UMARA.

March Full Moon
(A Ray II Invocation)

Part I

O Buddha Fire
And the Flame of Christ

O Buddha Fire of Divinest Wisdom,
O sweetest Light of the Flame of Christ:

A precious peace into me pour.

The princely person of me protect.
The penitent past of me purify.

The whole earth of me transmute.
The whole world of me transform.
The whole cosmos of me transfigure.

Around me flow now,
O Buddha Fire of Divinest Wisdom.

Let there shine forth through me, only
The sweetest Light of the Flame of Christ.

Part II

O Maitreyic Christ

Let me be carried safe upon wings of Grace,
Into the core of the Burning White Furnace.

Let the stars of the Heavenly LORD take their place,
Subdued and blissfully tranquil around my head;

And let the planets be laid to natural rest, forlornly
Naked, upon my Bodhisattva-sworn shoulders.

Let me be taken, open of palm and enlightened
Into the very Heart of perdurable Compassion.

Let my eyes glow forth spontaneously true,
The fiery look of Love, of the Maitreyic Christ.

May the Moon's Luminescent Manna fall into my palms,
And help fulfill the wondrous work of the Buddha Maitreya.

And may its Lucent Light bless and serve with coruscating
Compassion, all the sentient creatures of God's Creation.

April Full Moon

Christ, the Bodhisattva Maitreya

The Maitreya Buddha, the Imad Mahdi, the Kalki Avatar, the Bodhisattva, the Messiah, the Mirokou Bosatsu, the Divine Adventurer, and the Promised Saviour… different names according to different religions, and all describing the same person who has come once more to guide mankind.

The Bodhisattva Maitreya

Rays 2 3 1

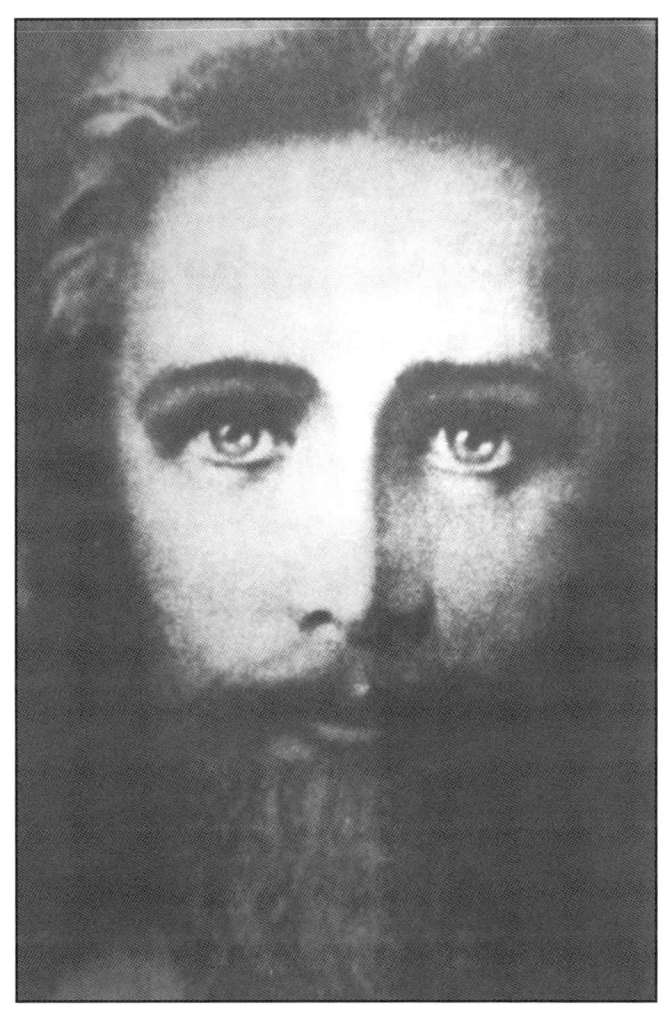

The Living, Loving Christ
The Greatest of the Lights

April or Easter, Full Moon

Part I

(The Risen, Living Christ)

By the Light of the April Full Moon, the Christ of Long ago, sacrificed His Life, and man sorely wept,

By the (same) Light of the Easter Full Moon, the resurrected Christ, calls today to one and all, both Great and small, to share His Life in the Perfect Plenitude and Caring, of the Father's JOY,

How the Christ Star Shines, scintillatingly bright upon the Earth's Easter Feast, invoking forth all the best of Love, and Light, and Life,
Coming down from Heaven's Silver Orb, in gentle blessing Rays to the realm of Man,

O Easter Moon, Full Bright and saturated to the round, with the Light of the risen, Living Christ:

Bring down silver-gleaming upon Thy Spirit Beams, the Divine Descent of the Seventh Ray, into the ready Mind of Humanity.

Bring down to us the New Order, the New Ceremonies, the New Vision and the Modern Modes of the Aquarian Teaching of the Great White Way.

Bring down to us, (upon Thy beams), the bloom and the blossom of LOVE, and the full-blown FIRE of He Who is called the Bodhisattva MAITREYA.

Bring down to us presently, the Living Christ, the Lord Who was the past well-Loved and renowned Krishna Avatar, and Whom the Buddha Himself has always known, without beginning and without end, as His own Sweetly-Beloved Brother.

Bring down to us the acting Head of Holy Hierarchy, the Sweet Sagacious Savior, the Kalki Avatar and the Perfect Preceptor of the Divine Plan for Man.

O Easter Moon, Full-Bright and saturated to the round with the Glorified Light of the risen, Living Christ:

Let the Nature of God Imminently be known as Love; be seen as Love; be heard as Love, and be lived as Love.

Let this LOVE-Blessing be compassionately channeled to the World, through the Saving Grace of Lord Maitreya's mighty Heart,

Let the Nature of Man through his Will-to-Good, swirl, surge and swell; and forthwith ascend in Spirit to the very peak of Goodwill.

Let Right Relations ultimately reign amongst men, and let this Blessing of Deliverance tenderly fall to earth, from Regal Maitreya's mighty shoulders,

Let the Nature of the Planet, shrug off abidingly, all struggle and strife; and let there spread from land to land,

The boon of Peace, and Beauty, and Truth; Lightly-upheld and Lightly-sent in Blessing, from Maitreya's mighty Palm,

O Easter Moon, Full-Bright and Saturated to the round, with the Light of the risen, Living Christ:

May Maitreya's mighty Presence, impress Itself upon Thy descending beams, and penetrate deep, the Cove of each man's Heart.

May Maitreya's mighty Presence, impress Itself upon Thy descending beams, and penetrate deep, The Tabernacle of each man's Mind.

May Maitreya's mighty Presence, impress Itself upon Thy descending beams, and penetrate deep, the Cathedral of each man's Soul.

May the risen Maitreyic Christ take hold of the Whole World, and Bless Mankind with the profound Purpose of His Expressed Love.

May the risen, Living Christ, stand with Mankind today and help Us to Resurrect, in what will become, the Greatest of the Lights.

Part II

(The Living, Loving Christ)

Know that the Transcendent Lord has taken Present Birth upon earth through the medium of the Saving Sonship, Revealed Radiance, and Real Resurrection of the Transfigured Christ.

Know that the chosen Queen Consort, the transforming Earth, has laid her mattering Body down, naked to the Test —

A velded Virgin of Variegated Veils on the vivifying verge of receiving Her recognized Savior Lord — the Living, Loving Christ.

Significantly, on this Easter Full Moon, may the Living, Loving Christ move within us as the Song of Songs.

May the Living, Loving Christ also move within us as the mushrooming Mantra of expansive Peace.

May the Living, Loving Christ move within us as the Crystalline Hope and Golden Glory of a whole New Age in Candescent Coming.

Part III

(Litany to Buddha MAITREYA, the Christ Lord)

O Easter Full Moon, beam down to us gloriously gleaming upon Thy Spirit Stream: Him Who is called "The Teacher of Men and Angels alike".

Beam down to us, *gloriously gleaming*: "The Matchless MASTER of all Masters."

Beam down to us, *gloriously gleaming*: "The reigning Regent of the Forces of Light."

Beam down to us, *gloriously gleaming*: "The Winsome Wielder of the Sword of Spirit."

Beam down to us, *gloriously gleaming*: "The Commander-in-Chief of the Armies of the LORD."

Beam down to us, *gloriously gleaming*: "The present Incarnate Buddha, or the Great (Golden) Godman, and Supreme Liege of His People."

Beam down to us, *gloriously gleaming*: "The Prototypal Spiritual Friend of Mankind."

Beam down to us, *gloriously gleaming*: "The Embodied (Enlightened) Expression of the Love of God."

Beam down to us, *gloriously gleaming*: "The Holy Healer of the World and the Sweet Savior Himself."

Beam down to us, *gloriously gleaming*: "The Radiant Prince of Peace and the Blazing Sower of (Future) Harmony."

Beam down to us, *gloriously gleaming*: "The August and Exhaled Emissary, or the Esteemed Envoy of SHAMBALLA."

Beam down to us, *gloriously gleaming*: "The Substantial SOUL of All Reality, or the Manifest Essence of All Emptiness."

Beam down to us, *gloriously gleaming*: "The Eldest in a Great Family, or Divine Lineage of Brothers."

Beam down to us, *gloriously gleaming*: "The Divine Director of the KINGDOM OF HEAVEN", by the authority of the Numinous Name of SANAT KUMARA,

And his Council Chamber of Kumara Buddhas in SHAMBALLA, for the Greater Glory and Evolutionary unfoldment of our (Future Sacred), Planetary Heart.

So Be It with humble Gratitude
on this Sacred Day of the
Full Moon Festival of EASTER

__ __ ____
D   M   Y

The Buddha

The Buddha, Bringer of Light and Wisdom. He presently operates from the most elevated etheric center of our planet, Shamballa. He acts as the Divine Intermediary between Shamballa and Hierarchy, and works intimately with Maitreya the CHRIST, in order to assist Him in His present Great Mission.

The 'Wesak' Full Moon

Rays 2
      3
      1

The Buddha
Incarnation of Compassionate Wisdom

Buddha Full Moon
(The Month of May Invocation)

The Buddha closed his eyes upon the month of May and the universes of man grew quiet.
The Buddha rose his hand in Spiritual Blessing and Peace penetrated deep the human heart;
The Buddha smiled serene an Inner Smile and the galaxy stopped and glowed in Wonder;
The Buddha prayed a silent Prayer and the Bodhisattvas in TUSHITA Heaven bent low an ear to Hear.

O Buddha, I would cease to breathe without Thine eyes closing softly upon me, Thy Grace Anointing.
O Buddha, I would cease to breathe without Thy hand guiding gently the Staff of my life's Path;
O Buddha, I would cease to breathe without Thy smile healing me whole and well, from the depths, Within;
O Buddha, I would cease to breathe without Thy prayer permeating the O.M., throughout the Sound of my Being.

O Buddha O, the world would break down and weep, without Thy Light Illuminating its thorny Way.
O Buddha O, the world would break down and wail, without Thy Wisdom directing its manifest Destiny;
O Buddha O, the world would break down and mourn, without Thy Compassion ministering balm to its Suffering;
O Buddha O, the world would break down and die, without Thy Love englobing it Whole in arms of Light.

The world and I, O Buddha, in this month of May and the moon at full glow, ask Thee humbly for Thy great yearly Blessing.
The world and I, O Buddha, in this month of May and the moon at full glow, impetrate Thee for Thy Primordial Awareness and the Amrita of Thy Empty Essence;
The world and I, O Buddha, in this month of May and the moon at full glow, petition Thee for both Thy Peace and Harmony to hold sway upon Earth;
The world and I, O Buddha, in this month of May and the moon at full glow, request of Thee the Equanimity of Adi Samantabhadra's 'All-Goodness'.

May Thy Tranquil Spirit enter into me, O Buddha, and fill me to the cup's edge with the Cool Fire of Thy Great Life.

May Thy Calm Mind enter the Earth, O Buddha, and fill its Chalice to the cusp with a Dancing Desire for the DIVINE.

Section II

131 Names of Buddha - Wesak

"OM NÂMO BUDDHÂYA!" (once, by 7 different people)

« Muni, Muni, Mahamuni, Shakyamuni Bodhi Svaha » (x3) (everyone)

"We bow to you respectfully, O Buddha
O Great Guide and Teacher of both gods and men
O Great Elder Brother of that Great Lover of Humanity, the Christ.

We bow to you respectfully, O Buddha, O Blessed One,
Spiritual marvel of the three thousand worlds of Creation,
Honored in Heaven as on Earth, honored as the precious, priceless *Jewel of the Heart*."

WE BOW TO YOU RESPECTFULLY, O BUDDHA

1. O Lord of Primal Energy's Dynamic Wisdom *(in Shamballa)*
2. O Bringer of Spiritual Light *(to Hierarchy)*
3. O Revealer of Divine Purpose *(in the world of men)*
4. O Leader of the Forces of Enlightenment *(everywhere)*
5. O Divine Emissary *(of the True Light of the World)*
6. O Bringer of Illumination *(to the world of beings)*
7. O Illuminator of (the) Mind *(of men)*
8. O Releaser of Light *(upon Humanity)*
9. O Lord of Compassion and Understanding
10. O Incomparable Conveyor of Enlightenment *(to our Planet)*

Bodhi Swaha (x3)

WE BOW TO YOU RESPECTFULLY, O BUDDHA

11. O Living Lord of Light
12. O Peerless Palm of Peace
13. O Founder of the Path of Illumination
14. O Formulator of the Light of Truth
15. O Instigator of the Way of Release *(from Rebirth)*
16. O High Road to Buddhi, *(and Awakening)*
17. O Divine Way *(of Dispassion, Detachment and Discrimination)*
18. O Divine Dissipater *(of the Game of Glamour)*
19. O Deliverer of the Human Spirit *(from the muck of matter and the e-motion of astrality)*
20. O Lord of Liberation *(through the means of Light)*, ("O Light of Liberation!")

Bodhi Swaha (x3)

WE BOW TO YOU RESPECTFULLY, O BUDDHA

21. O Inspiriter *(of Enlightened Mass Intent)*
22. O Sun of the East and True Son of Mind
23. O Knower of the Mind of BRAHMA
24. O Hidden Infuser of Light *(into the world)*
25. O Exalted Exhilarator *(of SELF Intent)*
26. O Venerable Lord of Vaisaka, (Wesak)
27. O Cosmic Channel *(of World-Contact)*
28. O Disperser of Darkness *(from the Hearts and Souls of men)*
29. O Occult Agent *(of the Planetary Luminary)*
30. O Catalyzer of Knowledge (transformed *into Sagacity*)

Bodhi Swaha (x3)

WE BOW TO YOU RESPECTFULLY, O BUDDHA

31. O Skilled Healer *(of all conditions of pain, suffering, & sickness)*
32. O Supreme Orator *(of Omniscient Speech)* & *(the 'Bright of All Orators')*
33. O Calm Conveyor *(of the Bliss of Amrita)*
34. O Living Lamp *(of the Whole World)*
35. O Crystal Mirror *(of Absolute Stillness, Emptiness, & Clarity)*
36. O Luminous Door *(of Divine Dharma)*
37. O Infinite Inspiriter *(of the Lighted Path)*
38. O Grand Lotus *(of Great Men)*
39. O Fearless Lion *(of Fearless Teachers)*
40. O Bright Full Moon *(over Mount Meru)*

Bodhi Swaha (x3)

WE BOW TO YOU RESPECTFULLY, O BUDDHA

41. O Magnificent Moon *(of Shining Merit)*
42. O Great King *(of Noble Bodhisattvas)*
43. O Blessed One *(Who walked the Earth and sowed seeds of Light)*
44. O Noble Being *(of the Four Noble Truths)*
45. O Deliverer of the World *(from the realms of Desire)*
46. O Substractor of Suffering *(from the Stage of this World)*
47. O Diamond (or Vajra) Heart *(of Profound Divinity)*
48. O Divine Dispeller *(of Ignoble Ignorance)*
49. O Illuminator of Men *(seeking Enlightenment)*
50. O Master of Men *(who have become empty Chalices of Compassion)*

Bodhi Swaha (x3)

May Full Moon

WE BOW TO YOU RESPECTFULLY, O BUDDHA

51. O Spotless One *(of Purest Divinity)*
52. O Wise One *(of Quiet Being & Superior Attainment)*
53. O Perfect Jewel *(reflecting the Light of Life)*
54. O Silver Chalice *(of Pure Emptiness)*
55. O Masterful Rotator *(of the Wheel of Dharma)*
56. O Master Planter *(of Buddha-fields)*
57. O Divine Destroyer *(of Demonic Domains)*
58. O King of Physicians *(and Prescriber of Complete Deliverance)*
59. O Angelic Charioteer *(of Lost Beings)*
60. O Blissful Ocean *(of Calm Magnificence and Supreme Virtue)*

Bodhi Swaha (x3)

WE BOW TO YOU RESPECTFULLY, O BUDDHA

61. O Great Bodhi Tree *(for both Seekers and Bodhisattvas alike)*
62. O Merciful Muni of Mount Meru *(with Compassion for countless sentient beings)*
63. O Splendiferous Selflessness *(shouldering a Heavy Load)*
64. O White Sky-Elephant *(of Ineffable Effulgence)*
65. O Heavenly Hero *(of men and gods alike)*
66. O Spiritual Master *(of the Sweetest Voice)*
67. O Absolute Delight *(of the Divine Dharma)*
68. O Great Guide *(across the Great Divide... of Emptiness)*
69. O Definite Destroyer *(of Disease and Death)*
70. O Victorious One *(over samsaric desire and the fettering passions)*

Bodhi Swaha (x3)

WE BOW TO YOU RESPECTFULLY, O BUDDHA

71. O Peerless Protector *(of the young, the old, the weak, and the sick)*
72. O Magnificent Muni *(of the Full Moon of May)*
73. O Holy Purifier *(of the Eye of Dharma)*
74. O Tall Tathagata *(of the Essential Truth)*
75. O Impeccable Teacher *(and most Pure Possessor of the stainless Doctrine of Dharma)*
76. O Blessed One Who Shines Equally *(upon all beings in the three worlds of Creation)*
77. O Paramount Producer *(of the Path to Nirvana)*
78. O Numinous Navigator *(who has crossed the sea of Rebirth and arrived at the Other Shore)*
79. O Beatific Bhikshu *(who broke the shell of the Egg of Ignorance)*
80. O Regal One *(Bearer of the Precious Parasol of Purusha)*

Bodhi Swaha (x3)

WE BOW TO YOU RESPECTFULLY, O BUDDHA

81. O Divine Walker *(of the seas of Mercy)*
82. O Exalted River of *(Omniscient Knowledge)*
83. O Smasher *(of Doubt, Discouragement and Despair)*
84. O Divine Diadem *(of the Undefiled and Pure Void)*
85. O Thrice Blessed One *(endowed with the "seven sparkling facets" of Enlightenment)*
86. O All-Envisioning One *(of Transcendental Wisdom)*
87. O Most Splendorous One *(Whom everyone everywhere, never tires of looking at and Meditating upon)*
88. O Supreme Shedder of Light *(bringing instant Joy and Delight to all beings)*
89. O Display of the Glory *(of SELF as Light and more Light)*
90. O Peerless Paragon of Prajna *(for both Learners and Knowers of the "Prajnaparamita-sutra")*

Bodhi Swaha (x3)

WE BOW TO YOU RESPECTFULLY, O BUDDHA

91. O Undisputed Leader *(directing men out of Darkness into Light)*
92. O Uncontested Dispeller *(of all questions)* and Calm Clarifier *(of confusion)*
93. O Tranquil One Who is likened to Candra *(with the most complete Calm and composed Demeanor)*
94. O Insightful Interpreter *(of the Master Mandala of Universal MIND)*
95. O Scintillating Cintamani of the Heavens *(the most precious and kingly of Spiritual Gems)*
96. O Victorious One, "Jyoti Jina" *(Who Has Crossed Heaven upon the Cross of Renunciation and Who has plunged as a Liberating Light into the very pits of Hell itself)*
97. O Blessed One *(with the Blue-Black Eyes of Compassionate Benevolence)*
98. O Great Comforter, Refuge and Aid *(for the plenary World of Samsara)*
99. O Great Vajra Bearer *(of All-Knowingness)*
100. O Laughing Lama and Lotus of Light *(reverberating as the Shimmering OM, in all lokas)*

Bodhi Swaha (x3)

WE BOW TO YOU RESPECTFULLY, O BUDDHA

101. O Most Holy Form *(of Unblemished and Radiant BEAUTY)*
102. O Pristine Mind *(of the Perfect View)*
103. O Holder of the Great Banner *(of Illimitableness and Immeasurableness)*
104. O Holy One *(beyond reproach in body, speech, and mind)*
105. O Divine Dweller *(in Desirelessness, Imponderability and Emptiness)*
106. O Beholder *(of the Terminal Truth)*
107. O Livingness *(of Absolute Being & Absolute Absence)*
108. O Inimitable Expression *(of Thatness, Isness, & Ain'tness)*
109. O Effulgent Exemplar *(of Empty Essence)*
110. O Bright Aborter *(of Cause, Condition(ality), & Circumstance)*

Bodhi Swaha (x3)

WE BOW TO YOU RESPECTFULLY, O BUDDHA

111. O Celestial Purpose *(with the Strong Stride of Shamballa's Will)*
112. O Rainbow Bridge *(to the Rainbow Body)*
113. O Great Tree of the Sky *(mature with the flowers of Knowledge and the fruit of Emancipation)*
114. O Peerless Possessor *(of true Discipline and Steadfast Sadhana upon the Path of Dharma)*
115. O Fairest Udumbara Flower *(of rare Bloom, rare Sight, & rare Purity)*
116. O Golden One of the Gods *(sprinkling to all corners of Creation the "gold dust" of bodhicitta, maitri, patience, and generosity)*
117. O Sublime Strength of Narayana *(gently subduing Mara & all opponents)*
118. O Blessed One of Broad Being *(bringing all things, time, and people into correct accord and right relation)*
119. O Matchless Meditator *(of all that is Clear Awareness)*
120. O Happiness of Heaven *(always giving and never sad to Share, with all beings of Sentiency)*

Bodhi Swaha (x3)

The Science of Full Moon Invocations

WE BOW TO YOU RESPECTFULLY, O BUDDHA

121. O Venerable Vahana *(of the Essence of Dharmakaya, Sun of Sambhogakaya, & Rays of Nirmanakaya)*
122. O Unmitigated Channel *(of Transmission & Transformation used by the Ancient Gods of the Holy Tantras)*
123. O Light Bearer from On High *(and Incomparable Trainer of those who choose to tread the Splendrous Empty Way)*
124. O Blessed Builder of Light *(upon the Lights of the past, present & future Buddhas and Bodhisattvas)*
125. O Profound Compassionate Solutioner *(of sorrow, suffering, and misery)*
126. O Thousand Eyes of Long Sight *(seeing without obstruction, nor difficulty, the karma, death and rebirth of 60 billion beings in all cycles of Samsara)*
127. O Eye of Awareness *(clearly perceiving Perfection beyond all defilement, without beginning & without end)*
128. O Fine One of the Fearless Affirmation that "I Am the Light of the World" *(which cannot be opposed by Heaven, nor Earth)*
129. O Great Abider in Equanimity, Non-Duality and Tranquillity — *(calming the restlessness of the mind and attenuating the agitation within all beings)*
130. O Matchless Samadhi of Non-Diminishing Effortless Effort *(in regard to the BLESSED ONE, Who will not quit, nor turn back until complete Liberation has been attained by one and all)*

(The 7 individuals representing each Ray are to recite this together)

131. O Maitreya Buddha, even before the Christ came and shared your Name, *deep, subtle, and difficult was the Divine Dharma to see; and seemingly no end was there in sight to Thy Turning of the Wheel in response to "The new Teachings of Wisdom"... as the Wheel of Dharma even today, rotates by Natural Law into Maitreya the Christ's, peerless Demonstration of Perfection... through the (integrated) personality's Discipline of LOVE".*

Section III

The Buddha-to-Christ Transmission (Upon Ray I)

Of burnt Gold is the Buddha's entire Being
As He radiates His abundant Blessings
Upon the month of May.

He cosmically stirs and stretches the magnetic muscles
Of His great expansive spatial Expression;
As He *Will*-fully pulls into Himself,
A pulsating Power of the deepest Profundity,
And of the purest Emptiness of Essence.

The Shamballa Power, tornado-gathers mysteriously around Him,
And seethes about His great heaving Breast,
Until it grows inscrutably Stilled.

Subsequently, steady upon a secret Sensa Sign,
The Blessed One whooshes the Whole Phenomenal process
Outwards, as the armipotent Principle of Dynamic Light —

Straight to the expectant CHRIST's great receiving Mitt
Of Divinely Tranquil Consciousness.

And quick-as-a-snap, the mighty MAITREYA Himself,
In His turn, relays the All of It,
To His Seven Cosmic Catchers,

Upon the Seven Bases of the ONE CREATOR's
Creative Expression of the WILL-TO-GOOD
Upon the Great Wheel of LIFE.

Section IV

Transmission of the Will-To-Good
upon the Sub-Rays (of Ray I) to Humanity

(Recited by the respective 7 Rays)

R1 As a result, the Will-to-Power will energize the Lord's disciples everywhere, and through High Initiates will guide the new Visionary Politics of man.

R2 As a result, the Will-to-Love will override all mistrust and hate in everyone and truly inaugurate the Aquarian Brotherhood of the New Golden Millennium of man.

R3 As a result, the Will-to-Action will activate a refreshing world of equal Sharing, and instigate a fresh economy for the greater well-being of proletariat man.

R4 As a result, the Will-to-Harmony will put all men in right relationship to each other and to the Logoic Lord, and there will be renewed negotiation and a re-inspired cooperation and harmony amongst men.

R5 As a result, the Will-to-Know will take a central place in the hearts of men and will quicken forth throughout Humanity an authentic hunger for real Knowledge and synthetic Wisdom.

R6 As a result, the Will-to-Devotion will take firm root in Hierarchical Ideals through the pursuit of Spiritual Happiness via the goal of Self-Emptied Realization; and the new (Ideals) of Innate Infinity and Inherent Immortality, (no beginning, no end), will be persistently shown as being *basic* and *essential* to the very Nature of man.

R7 As a result, the Will-to-Order will organize itself in Hierarchical *ceremony* and *enterprise*; and a Golden Order of the new Aquarian Man will know the Impersonal Inspiration of the (One) LORD of Shamballa, through the personal demonstration of CHRIST's recognized BUDDHAHOOD.

(Recited by the 13 people together)

"OM Namo Buddhaya" (x3)
"OM Namo Maitreya" (x3)

The Buddha Maitreya, the Christ

The Kalki Avatar, the present Buddha Bodhisattva, the Messiah, the the Divine Adventurer, the Promised Saviour, Imad Mahdi, and the Mirokou Bosatsu... different names according to different religions, and all describing the same Initiator who has come once more to Instruct and guide Mankind.

The Buddha Maitreya, The Christ

Rays 3
      2
      1

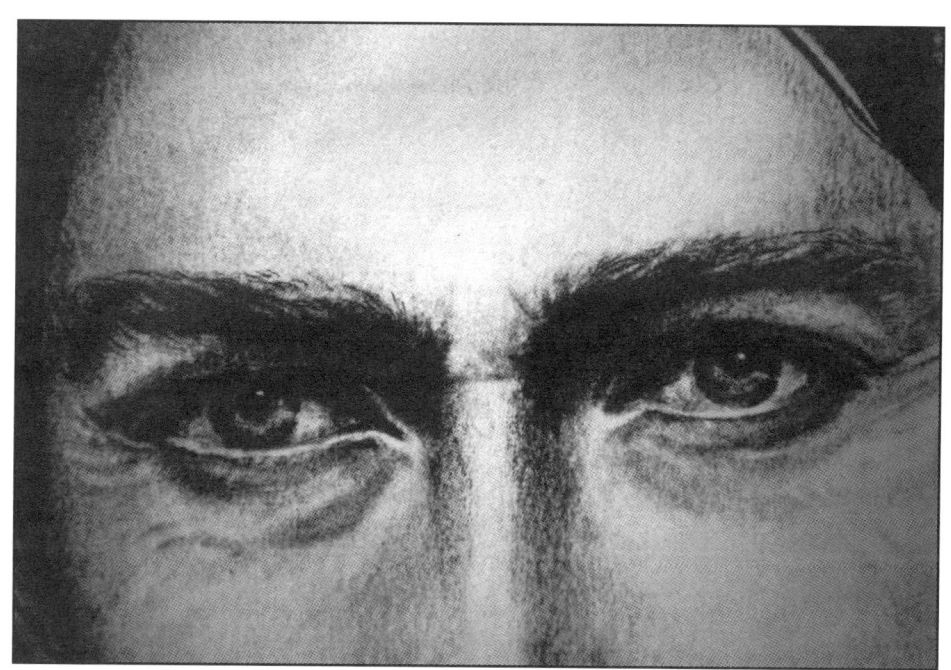

'He Comes,
the Love of God empowering His eyes,
the Father's Will investing His Heart'

He Comes
(The June Full-Moon Invocation)

He comes,
the Love of God empowering His eyes, the Father's Will investing His Heart.

He comes,
as the Nourisher of the Little Ones, as the Dispenser of the Waters of Life.

He comes,
in this New Age of Awakening to lead the Aquarian Sons of Liberation and Light.

He comes,
in this Diamond Age at dawn, to Initiate the sons of men into their long period of Golden Peace.

He comes,
to quell the noise, the discord, and the disharmony of Humanity's incessant quarrelling.

He comes,
to take away the hate, abate the rancor, and to appease the torment of men's minds.

He comes,
to still the competing breath; and to stay the strife that is so rife among brothers, among nations, among labor, among sects, and religions, and churches, and the young, and the elders.

He comes,
to patch the distrust running deep; to smooth the chaos of conflicting concerns.

He comes,
to halt the hand of war; to outcast the critical clamors; to erase the muck of spiteful talk.

He comes,
to relieve the sorrow, and the struggle; to soothe the suffering selves, and to heal the cancerous wounds.

June Full Moon

HE COMES,
to dispel the dismal dark; to lift and lighten the burden, and to tear down the veil of false separation.

HE COMES,
with the Light of the Buddha upon His breath; with the Wisdom of the Buddha inspiring His Soul.

HE COMES,
the Buddha standing full-support behind Him, with the month of May humbly conceding way to greet the coming Great Task of June's re-shaping of the Souls of the sons of men into the Light and the Spiritual Sight of their becoming conscious creative Sons of God.

HE COMES,
to bind and blend powerfully the Love-Light of the East to the Scientific-Light of the West.

HE COMES,
to establish a Spiritual Solidarity of Supernal Sorts among the sisters and the brothers, and the sons and the daughters of men and women and children everywhere serving God, and each other.

HE COMES,
to span forth the mystical age of passing Pisces, fusing it full-blown on fire and fully-Realized into the cool Ceremonial Mind and Invocative Rhythm of the New Aquarian Dispensation.

HE COMES,
as God-Transcendental-Descending, the Master of Masters, the Teacher of both men and Angels alike.

HE COMES,
as the Divinely-Appointed, acting, living Head of the Spiritual Hierarchy of this planetary scheme.

HE COMES,
to stir Awake the reality of God-Immanent, and to install It in the Life of all manner of men, and beast, and things.

HE COMES TO SAY:

"Thou are not alone.
Follow me out of the storm.
Follow me into the Haven of God's kingdom,
whose time has now come to be firmly and publically established upon earth.
Come forth and build with Me Humanity's Temple of Peace and Goodwill.

Let all good men, and women, rise in the new Religion of Light.
Go forth, and Sound the Trumpet that will unite all disparate faiths under the Divine Banner of the One God of Love."

"Let the Will of the Father be Revealed.
Let the sons of men recognize the Maitreyic Christ, That Son of God, Whom I Am."

"The Christ I Am, the Maitreya Buddha I Am.
The Imam Mahdi I Am, the Kalki Avatar I Am.
I Am That I Am: the Bodhisattva, the Messiah, the Mirokou Bosatsu, the Divine Adventurer, the Promised Savior… come again to lead man and His little ones into the enduring Joy of the Victory of Light."

HE COMES,
as a simple man, with a simple Teaching, as you and I seeing the Self-as-ONE: the Sons and Daughters of the Same God of Truth.

HE COMES,
as a humble man, with a practical penchant for Giving: asking that each give what he can in the sharing of the Wealth.

HE COMES,
as a kind man, in treating Christly the heart of all.

HE COMES,
as a medicine man, in helping to heal the pain of the past, and to wash away the scars of matter.

HE COMES,
as a far-Seeing man, in the laying down of just government; and the stabilization of right Human Relations among Nations.

HE COMES,
as a Knower of God, to recruit and to Initiate men into the secret, sacred Knowing of His Will.

June Full Moon

HE COMES,
to help men gain access to that great Apostolic Succession of the Knowers-of-God, whose Body Harmonious-and-Luminous, consist of those Individuals and groups of Servers whose footsteps of Light leave a fiery, Living Path for those who Love but only Him.

HE COMES,
not as a man of sorrow, not to be persecuted, not to be crucified;
but as a Joyful man of fully-Realized Divinity to teach Initiation, Goodwill, and Right Relation.

HE COMES,
not alone, but with his Ashram of Masters trained-in-God, and taken from all human-ranks; to restore the Arcane Schools, the Ancient Landmarks, the Occult Sciences, the Sacred-Spiritual Arts.

HE COMES,
as the Point-within-the-Triangle, pouring forth over the masses the energies of Love, of Light, and Greater Life.

HE COMES,
with the AVATAR OF PEACE tranquilly descending.

HE COMES,
flowing as the wind from the Himalayan shoulders of this Spirit of Peace.

HE COMES,
the AVATAR OF SYNTHESIS rising below Him, upholding Him as the Divine Distributor of that Great Triune Energy... of Love, and Wisdom, and Will... Immortally-fused into the One-Light-by-Him-held, and Blessing all, Powerfully.

HE COMES,
to carry the Equation of God to Enlightened man.

HE COMES,
as the Prince of Perfect Peace and the Lord of Perfect Equanimity.

HE COMES,
to ordain and to sustain the Aquarian Order of the New World Servers of One Humanity, that is, of those Adepts and Initiates, disciples and aspirants of Hierarchy, willing to incarnate the Shamballa Plan of God upon earth.

HE COMES,
Yes, Maitreya comes as the Christ,
into the warm June of men's expectant hearts.

HE COMES,
The Christ is here presently among us.
Therefore, rise up ready and pure, O Pilgrim, to meet the wonderful summer of His Love.

Invoke Him sure into your lives, O Children of Earth:

"O Maitreya, take me to Thee in the burning of the night.
O Christ, my Love-Light, harken to the sun of my Soul pulling Thee irresistibly to me.
O Maitreyic Christ, let me be Thine in mind, and heart, and will; and let my Spirit shine with Only Thee.
O Maitreya-Buddha, O Christ, One and the Same; bless our planet land, and we do Love Thee in the Light.

Let mere belief in Thee give way to the Greater State of Knowing Thee in Thy humble Glory.
Let us with the fading of faith into the Bright of Knowledge, become one and all, keen Knowers-of-God.
Let us ray forth in our life the Blessing of a closer contact with the Divine privilege of Thy Physical Presence.
Let us be trained by Thee and Thine in the potent Science of Invocation, for Love, Liberation, and Greater Life.

Let us meet at the God Center of the Triangle where Humanity, Hierarchy and Shamballa resound 'United'; and let the Three-combined-into-the-ONE, command Evolution and proceed with the Work of creating Heaven upon Earth.

Let the Seekers of the Light and the Lovers of God from all sects, and religions, and churches recognize their Common Spirituality and their One Soul, and meet on the common platform of the New World Teacher.

Let the Christ be recognized, Transcendentally, by His striking deep the chord of God-Immanent within the hearts of men; and let the great wordless Cry, caught in the throat of a suffering humanity, be released into the Healing Hands of Hierarchy.

And let it be presently resolved by us, through the united Goodwill and Right Relations of the people of the beautiful planet Earth — that under the Divine Leadership of Maitreya the Christ, Lovingly sent to us in the June of man's evolution by God-the-Father,

Will, that man live together as One Soul, One Humanity, in the Golden Peace and Powerful Synthesis of the Aquarian Dispensation.

May the Christ in this month-of-June be therefore beside us, and within our Hearts at all times.

May His Love and Light at the moment of the Full Moon, stir within, the Will-to-Good and the Will-to-Know God — in every man, woman, and child, stirring even the lambent breast of the beast, and the Heartbeat of the Rock of Time."

Section Two
Festival of the June Full Moon

I
Transfer of the Buddha's "Vestures"

The Buddha vivifyingly bestows *His Vestures* to the new Lord of the Aquarian Age, Thereby doting the Christ with the Mind of the East and Heart of the West.

II
Release of Light into Love

In all lands men await His promised Coming and all hands are cupped as One Chalice, to welcome the Amrita of Love pouring down from Him as Light into the Heart of Humanity.

III
Closer Contact

A closer contact with the Celestial Chord of Love upon the beams of the Full Moon of June,

Is the Gracious Gift (given to us as men), honoring the Ascending Impulse of Human Aspiration conjoined to the Illumined Intent of Hierarchal Descent.

IV

Our Minds and Spirits United

As dedicated Disciples and as a representative group of the 'race of men', we now reach up with our minds and spirits united,

And fixate the Core of our Being upon the Waters of Life flowing as Love from the Heart of the Living Christ.

V

Abundant Are His Blessings

The Christ is one with His Brother Buddha as Maitreya, the Munificent Muni.

And abundant are His Blessings and bountiful is His Grace upon the present Aquarian Age of Golden Promise and restored Heavenly Dispensation.

VI

A Future Shining

The Radical Radiancy of the present True Hour of the descending Christ Light,
Rushes through the open gates of this June's Full Moon of magical empowerment,
Toward the Centre of the Race of Men and inundates the Consciousness of Mankind,
And spreads softly as Love all over Earth and throughout all of the created kingdoms.

And the Earth's Core itself shivers momentarily and sighs the release of a supernal Future Shining.

VII

Humanity, Hierarchy and Heaven, Rejoice!

O festive Full Moon of the Sixth Calender Month, on this the ____ day (of June), the year 2_ _ _.

Gather Occultly to Thy great Silvery Bosom the Goodwill of the people of our planet.

May Thou accomplish what has to be done, in order to bring our Heart, as dedicated "servers", into a closer contact and more responsive relation to the Will-to-Good, which ray forth mindfully our Elder Brothers of Spiritual Hierarchy, and of which MAITREYA the CHRIST is its Living Head and Holy Host of Ceremony.

May Humanity, Hierarchy and Heaven rejoice in their sharing together as beneficent complices in Compassion, a newly-found Occult Triumvirate, consciously distributing the Triple Creation Energy of Love-Wisdom, Intelligence and Power.

May the all of it unfold forward into our Common Future in ordered Right Relation, under the vigilant Vision of the Planetary Logos, SANAT KUMARA and the watchful Eye of the present CHRIST in Aquarius.

VIII

Who Is There to Resist?

Who indeed, is there on earth to resist the Sweetness of the Unifier of East and West?

Who indeed, anywhere, is there to resist the Radiant Releaser of Effulgent Energy upon the earth planet, and its pilgrim people?

Who indeed, is there anywhere to resist the Perfect Preceptor, who will establish the New Religion of LOVE-as-Light, to be liberated equally and Intelligently (amongst us)?

IX
Toward "Greater Life"

Who indeed is there, who can resist the Desire of the Divine One Who once said, some two-thousand years ago:

"I have come that you may have Life more abundantly!"

X
"The Great Invocation"

May the forces of the Resurrection, Reformation, Reconstruction and Restoration of the world,

Go forth under the Inspiriting Enlightenment of the Great Invocation, as it is to be powerfully pronounced by us presently upon the "CHRIST's Unique Occasion", on this festive day of the Full Moon of June, the ____ day of the Roman Calendar year, 2_ _ _ A.D.

[Recite together the "Great Invocation"]

July Full Moon

Rinpoche, Guru Tsongkapa

Manifestation of Manjushri, the Buddha of Transcendental Wisdom. Djé Tsongkhapa was a great Tibetan Master during the 14th and start of the 15th century. He was able to demonstrate through personal example, the progressive path to Enlightenment and how to realize the Buddhadharma through Mahayana Buddhism, the ordinary and secret Tantras, and other occult practices.

'Lotus of Light'

Rays 3
      2
      1

Rinpoche, Guru Tsongkapa

Full Moon of July
"Lotus of Light"

On this shining full moon of the month of July, I bow profoundly to my Guru,
And pray earnestly that I be held dearly as a disciple of worth in his holy sight.

O Lama Guru, who are insightful and wise in all works arcane and knowledges considered Divine,
And who, through an immaculate mind, impeccable behavior, and correct unerring speech,

Convey to all who beseech you, to teach simply the clear meaning of the High View,
And whose powerful Presence extends out in blessings of kindness, compassion and love for all.

I pray to you, my precious Guru, to bless my form and guide my steps unfailingly upon the Way.
Push me to listen to you always (attentively), and put into ready practice your least suggestion, or meanest counsel.

O, my Guru, I prostrate myself to you, who are ablaze with the Shining Light of full Enlightenment.
May I follow you faithfully and true, in your bright, clear reasoning and in your flawless armor of Wisdom.

And in spite of these dark, degenerate times of sin and ignorance, your amazing Mind minds only the Divine.
Let us, I pray, combat together and lay waste with great Compassion, the vast armies of the tempter, Mara.

You are my perfect Master, who with a motherly love, has sworn to free all sentient beings from base enslavement to Samsara.
Too many today, are inextricably caught in the passing show of phenomenal illusion, and in the sorry delusion of Desire's karmic sizzle.

Your name and fame matter not, my Master, but your Being what you are (Naught), simply brings pure delight to my sense of right.

Your beams of bright Emptiness detonate diaphanously as a radiant rainbow within my inner sky,
And my Mind is hoisted high and hangs heavenly harmonious in absolute, inconceivable awe.

July Full Moon

Your constant clarity and high inspiration, my Master, make me ponder the possibility of your Mind being One with the Noble Manjushri.

Your unique humbleness of person and dedicated discipline of unceasing service, universally tended toward the ultimate enlightenment and true happiness of others,

Make my body bend and 'bodhi swaha' in the joyful recognition of that minute, magnificent miracle which I have before me and hold preciously within my heart.

You, who are the self-effacing Lama Lord of the Undefiled Dharma, the pure Perfect Master, who is my unsullied Lotus of Light and Agent of Unalloyed Liberation.

July Full Moon
(Alternate Invocation)

O Guru Full Moon and Mother of the Seventh Month of life arising,

Suffuse softly through the emptied spaces of my mind, Thy Divine Light downward to me descending.

Whisper in dissolved tones of O.M., Thy secrets tonight to the full heart of my Love Listening.

Speak to me receiving the Immortal Word, and pierce my Soul glad-full of God-glistenings.

Sound strong the Eternal Gong under the Bodhi tree and send me swirling into the empty Silence of my Full-Awakening.

August Full Moon

Lord Krishna

Krishna is one of the most venerated divinities in all of India. He incarnated, (circa 3000 B.C.), in order to restore order on Earth at a time of materialistic spirituality and historic difficulty. His unique and exciting life story has created innumerable accounts of legends symbolizing all aspects of human spiritual aspiration and development. His teachings are partly presented in the Mahabharata and Bhagavad-Gita.

'Magical Moon of Krishna's Flute'

Rays 2
1
3

Gopal Govinda
Lovely Lord of Alluring Love

Magical Moon of Krishna's Flute
(August)

Om Sri Krishna, Beatific Lord of Boundless Beauty.
Om Sri Krishna, Adored and Adorned Lord of Divine Devotion.
Om Sri Krishna, Lovely Lord of Alluring Love.
Om Sri Krishna, Bright Lord of Sublime Bliss.

Om Sri Krishna, Gokula Lord of Gopal Govinda.
Om Sri Krishna, Supreme Lord of Self-Surrender.
Om Sri Krishna, Kindling Lord of Kinetic Kindness.
Om Sri Krishna, Peerless Lord of Pure Pleasure.

Om Sri Krishna, Cowherd Lord of Consoling Compassion.
Om Sri Krishna, Sovereign Lord of Sweet Sacrifice.
Om Sri Krishna, Saving Lord of Sisyphean Suffering.
Om Sri Krishna, Merciful Lord of Merciless Misery.

Om Sri Krishna, Resplendent Lord of Rapturous Rasa.
Om Sri Krishna, Govardhan Lord of (the) Guileless Gopis.
Om Sri Krishna, Vitthala Lord of Vivifying Vrindavan.*
Om Sri Krishna, Heroic Lord of the Holy Mahabharata.

Om Sri Krishna, Victorious Lord of Viers of Vishnu.
Om Sri Krishna, Primeval Lord of Pastoral Play.
Om Sri Krishna, Revered Lord of Radiant Radharani.
Om Sri Krishna, Bhagavan Lord of Buddha Blessings.
Om Sri Krishna, Humble Lord of Human Hearts.

OM SRI KRISHNA... as the Aeonian Lord, you say that:
"Life is Love and Bhakti is Bliss".

That these two principles make up the perennial Primordial Pulse, and Heavenly Humming Harmony of the Heart of Sadhana, or Yoke of Yoga;

*Vitthala is usually associated with the town of Pandharpur.

That these two, make up the non-prejudicial passionate Path which passes paradoxically dispassionate
Through the polar paradigms of Arjuna's disciple dilemma of duality and Divinity,
Of form and Formlessness, of relativity and Relationship, of diversity and Oneness.

Let us, henceforth, return to the Divine Dictum that 'Love is Life', enhanced by God, the Krishna Lord.

Let us reiterate the Divine Dictum which says that 'LOVE with Devotion, leads naturally to Ecstatic Bliss',
Upheld by Hare Krishna, the Dark-Blue Lord, as 'He magically steals the minds and hearts of all souls'.

Let us return, again and again, to the Supreme Formula, and know in our Heart of Hearts
That 'Life is LOVE and Love is LIFE', and listen closely to Sri Krishna's pure flute notes,
As His simple tune uplifts us ecstatically (eternal) into the field of essential Freedom.

Let all human contention and human conflict be resolved and resorbed as 'Simple LOVE',
And be mindfully directed to the Divine One in pure tones of the selfless Om.

O Krishna, O Gopala Govinda, of such Beauty, of such Simplicity, is the karmaless Middle Way.

Om Hari Hare Om
Gopal Govinda
(inbreath) (outbreath)

Section II

August Full Moon
(Alternate Invocation)

O August Presence,

O rounded, reflected O., of the Primal O.M.'s descent of Light, upon the globe.

Carry forth upon Thy rays and into all hearts, the Courage to lay down today, beyond the obvious, pale truth of words, the silent Golden Commentary of the 'Empty State of Mind'.

Let me live today the wisdom of enlightened surrender, linked to the Intelligent Activity of SHAMBALLA's Cosmic striking of the Primordial Sound.

Let me today become a magnified Holy Auric Glow, glorifying the Sacredness and Is-ness of the ongoing moment-to-moment platform of the Omnipotent's Omnipresent Being.

Let me disappear today from the small earth of me rotating around my navel; and let me sit quietly still and bowed, before the Hallowed Birth of Lord Krishna.

By the Blessed Light of the Full Moon of August, let me reach up a hand through the beauteous beams of its Illumined Face; and ask the High Father for the Compassion of His Full Pardon and the radiance of His Forgiving Grace.

Let it be that I incline my body presently in humble prostration, in recognition of the Supreme Godhead and the state of Genuine Godhood.

May it be upon this day of present Power and in the perfect clarity of the Moon's silver and shine,
That I pledge myself unreservedly to the 'ONE about Whom Naught Can Be Said'.

Let me, in Complete Understanding and Full Awakening, ply the integrated personality Henceforth, with the SOUL's Solar Shining, into an Absolute Sonship of Loving Service,

Coupled to the indomitable application, or Fiery Intent, of Heaven's High View,
And molded to the august, purpose of SANAT KUMARA's Plan for the world of man.

September Full Moon

Ganesha

Ganesha is perceived as the Hindu Divinity having Perfect Wisdom. He is also the Harbinger of Good Luck and the Blockbuster Lord who removes all obstacles. He is the Lord and Master of all in the world which obstructs, restrains, prevents or hinders, materially, but especially spiritually. He can remove all impediments and adverse circumstances upon our path to Self Realization. Nothing should be started without first invoking Ganesha and offering Him both veneration and salutations.

'Ganesh, The Peripatetic Pilgrim'

Rays 1 3 2

Ganesha,
The Lord of Beginnings

Ganesh Chaturthi
Full Moon of September

Part I
Introduction
Geographical Traveller

Pristinely born in the rubicund light of the morning's rising sun in the holy land of mystical India, charismatic Ganesha travelled northwards to Kashmir, Pakistan, and Afghanistan, and then crossed over into the middle east, into cosmopolitan Turkey and Iran.

Coetaneously, he went to Nepal and then took to tackling the 'top of the world', that is, the infamous Tibetan Plateau; and thereupon, moving eastwards and onwards, he peregrinated to Mongolia, China and Japan.

And rapidly covering ever more ground, he itinerated to Burma, Thailand, Cambodia, and Vietnam, and took the whole of Indonesia by storm; and trekking on, he quickly conquered Bali, Java, and Borneo by the sheer force of his magnetic personality.

Mysteriously navigating between the continents, he cryptically ended up in Mexico, and crossed the whole of Central America; and even set foot, and left his mark upon the inscrutable Easter Islands.

And all the way along, he went always upon the wayfaring, peripatetic back of Hindu mythical cosmogony and iconography… supporting, endorsing, and inspiring their essentially mystical modes and manner of consciousness exploration, mantric exhortation, and poetic, or literary forms of creation… as they are apt to be found and perused in the inspirational writings of the holy Vedas and venerable Puranas; and albeit, also can be studied in the multivariegated metaphysical and philosophical schools, which illustriously gave to the holy land of India its great golden age of enlightened thought.

Part II
Historic
Glory of Ganesha

Glory be to Ganesha and his droll pot-belly chock-full of poetry, power and protection.

Glory be to Ganesha, the transcendental, transreligious, transcultural, translingual, transnational, awesome, yet utterly accessible Deity, who is guide, father and friend to all men in all lands where true spirituality is practiced.

Born upon a sunbeam and bespoken as the enlightened Surya-Ganapati in Nepal,
And to his becoming Dantin Vakratunda Ganapati, a portentous aspect of Shiva himself, as found in the Holy Vedas,
And maturing in time, to turn himself into the legendary leader of the Puranic forestial ganas,
And to becoming concomitantly, the highly honored friend of the playful Himalayan yakshas.

To at last, not long afterwards, being universally recognized as the vimful, stout victor and undisputed master of the vengeful vinayaka gang of 280 horrific hoodlums, known metaphorically since the dawn of Indian mythology, as the unruly and fearsome 'obstacle busters', and the daring and most dreaded brotherhood of 'danger dispersers'.

And when all is said and done, and in direct evolutionary line with the gallant Garuda and the heroic Hanuman, Ganesha took his final birth with the epitomic parents who were at that time cognized as the supreme Cosmic Couple... that is, he was born to Primordial Parvati, the most perfect and practical Holy Mother, and to the ascetic Sri Shiva, the all-wise Primal Father of mystical yogic accomplishment and mythical spiritual warriorship.

Part III

Invocative Celebration of Nine Ganapati Mantric Pilgrimages Leading to GANESHA CHATURTHI

OM GAM OMKARA SVARUPA GANAPATI (Namah)

'I am the form and vehicle of the continuous sound of OM, reverberating in all worlds as TAT TVAM ASI.'

'That is, the you that thou art as form existence, are by your very manifest(ed) existence, inseparably linked to the glorious Unmanifest, as also, THAT!'

Form, (the Manifest) and Emptiness, (the Unmanifest) are, from beginless time, one and the same; and the mystical meeting of them as 'Ga' (i.e., 'gaja', elephant) and 'Na' (i.e., 'nara' man), sets free the mythical image of *Ganesha*, or 'Ganapati', the Divine Elephant-Man which is considered Self-born, within the consciousness of both deva and man.

I bow to thee 'O, Omkara Svarupa Ganapati'!

OM GAM VIGHNA GANAPATI (Namah)

O, Great Obstacle Remover of burnished gold with the beauteous sensual shades of Kama-Krishna, and of ten manipulative arms to help distract and obstruct all negativity and any evil, (directed toward us).

I bow to thee 'O, Vighna Ganapati'!

OM GAM HERAMBA GANAPATI (Namah)

O, Great Guardian and Powerful Protector of the weak, innocent, and harmless. Of five grim faces and ten strong arms, valiantly riding a Durga lion as his mount, doing *abaya* and *varada*, and actively protecting and compassionately blessing all, in mystical mudra.

I bow to thee, 'O, Heramba Ganapati'!

OM GAM MAHA (TANTRIKA) GANAPATI (Namah)

O, Great Tantrika of the three mystical eyes and crowned with the crescent moon, sensually sitting upon the feminine (tantric) triangle under the famous 'parajita' tree, with his Shakti-consort in amorous embrace upon his leela lap, and his delicately holding a voluptuous pomegranate, and with the ample juices of sweet sugarcane flowing freely; and with the whole atmosphere aglow with soul-stirring devotion and responsive love.

I bow to thee, 'O Maha (Tantrika) Ganapati'!

OM NRITYA GANAPATI (Namah)

O, Ecstatic Dancer of bright golden color doing spontaneous and joyful dancing steps under the 'kalpavriksha', or wish-fulfilling tree... with rings of precious and semi-precious stones on his fingers, and a graceful free hand to state, signal and sign his intricate dance motions... which mysteriously, and intimately connects him to his famous father, Shiva the Nataraja, the Imitable Lord of Dance.

I bow to thee, 'O Nritya Ganapati'!

OM GAM UCCHISTHA GANAPATI (Namah)

O, sensually-spiritual, beautiful blue form of Ganesha, sexually aroused and erotically potent, nude, but bejeweled, and playfully embracing his Shakti; and with his mindfully stringing forth sweet music from the veena; and his ardently doing powerful mantra upon the sacred mala, at the same time... O, most Holy Hotness and Sweet Dadi Cool!

Glory be to thee, O most potent Fertility God and Ishwara Erotica!

I bow to thee, 'O Ucchistha Ganapati'!

The Science of Full Moon Invocations

OM GAM YOGACHARA GANAPATI (Namah)

O, Yoga Lord of most beauteous body, bathing in the bedazzling tones of the rubicund rising sun. Manly-mantled art thou, in the subtle modulations of Indra's heavenly-blue robes. Postured thou art in the ascetic mudra of the classical yoga stance, and traditionally engirdled, as you heft ever so lightly the hoary staff of yog. And with a crystal prayer mala in hand and mindfully doing holy mantra, you appear awesome in thy divine demeanor and real in thy spiritual realization.

I bow to thee, 'O Yogachara Ganapati'!

OM GAM LAKSHMI GANAPATI (Namah) (Swaha)

O, Benign Bestower of Wealth and Wisdom, with his two enlightened wives, Siddhi, (Goddess of the power of Success, Achievement and Accomplishment), and Buddhi, (Goddess of Infinite Intelligence, Knowledge, and Prajna), sitting devotedly on each leela lap, and each holding forth divine blue lotuses in consummate loyalty to Him, the elemental Elephant Lord of pure white Luminosity.

I bow to thee, and humbly serve thee, 'O Lakshmi Ganapati'!

OM GAM CINTAMANI (THEUR) GANAPATI (Namah)

O, Great and Awesome Protector of the Precious Jewel of Mind, as being both enlightened and pure. Defender thou art of the precious Heart Jewel, or pure Diamond Drop which thou hast rescued from the Infinite Ocean of Ambrosial Consciousness.

Only because of such a magnificent moment of Manifest Compassion, are all names and forms now experienced as thy Self and the Self of every man, full to the brim of Blessed Bliss and the silent Essence of Emptiness.

I bow to thee, 'O Cintamani (Theur) Ganapati'!

In invoking an adequate closing note to Part III, let us recite:

The 'Golden Age' Ganapati Mantra

OM MAHA (AVATARA) GANAPATI (Namah)

O, Great Lord, fiercely standing, one big arm in the blessing 'abaya' mudra mode, the other powerful one brandishing a mace as a palladium against evil, and who, in coming modern times shall come to be known as the Great Avatara Incarnation of Ganesha, the Ultimate Peacemaker and Creator of the Revolutionary Cycle, or new evolutionary era predestined to become the 'Wisdom Kalpa', or 'Golden Age' of a mature humanity.

I bow to thee, 'O Maha (Avatara) Ganapati'!

Note: 'I bow to thee'... 'I' or 'we' are interchangeable.

Part IV

'OM Srinathadi Guru Ganapatim Namah'
'OM, salutations to the Primordial Guru, to Ganapati, (I bow)'

In closing, know this of Sri Ganapati and be appeased, "OM Shantih, Shantih, Shantihi!"

1. He is the singer, the singing, and the song of OM.

2. He is God, Guru, and the Self.

3. He is Lord Agni and the Kundalini Shakti; he is the sparkling fire of the Divine and the kindling warmth of the mundane.

4. He is the fiery pot into which the will of sacrifice is poured and all experience stirred, until the molten gold of pure consciousness is achieved.

5. He is the first primal Ishwara of the Divine Pantheon and the chela's 'adi ishta devata', first personal deity.

6. He is the fiery seed of the soul, the Divine spark, or the Self, within.

7. He is the pot-bellied protector of the meek and weak, and in his western form as Saint Nicolas, the generous boon-giver in all lands of the occidental west.

8. He is the Divine Impulse of all Dance, the Divine Dancer, issued from the Bliss of Parvati awhirl to Nataraja Shiva's cosmic boogie.

9. He is the sagacious start of every activity and it is he alone who blesses every new beginning.

10. He is the bodacious bearer of anyone going over the murky waters of life, and of anyone's personality being tossed about by the transformative waves of disturbing change.

11. He is the Lord and friend of all beings, whether human or divine.

12. He is to be experienced immediately and imminently, visibly or invisibly. Think of him, and he is there; pray to him, and he is all ears.

13. Every second is his (own) time and every moment his present manifestation in space.

14. He is in every man's mind minding his mind and compassionately caring for all who are of pure heart.

15. He is the palpable, perfumed presence of Hereness, ready to guide at all times.

16. He is the Loving Lord of all learning, culture and the arts; and of all which encourages and concerns, the subtler skills and finer aptitudes of man.

17. He is shakti, buddhi, vidya jnana and prajna all in One Divine Breath.

18. He is the potent Lord of Arousal and the gentle Awakener of Feminine Fertility.

19. He is the 'svayambhu', the Self-born manifesting as the Adamantine Mind within the illumined Cave of the Heart.

20. He is the Real Self and not mere supposition that the self is real. Actively seek him out and listen in silence for His Primeval OM sound.

21. He is the "repeller of revilers" and "cleaver of the clouds of ignorance" and "knower of the unknowable", as well as the illimitable infuser of light into the mind, and the disperser of darkness within the sadhaka's sadhana, and his practice of meditation.

22. He is the divine drop of the Soundless One being just now heard, as well as the divine drop of Light resounding everywhere vivified in joyful reverberation upon the waters of Life.

23. He is the great big belly of absolute mystery occulting and harboring the Quintessential Seed of the Manifest.

24. He is the Hymn of the Holy, thundering silently from the Circle of Life to the Circle of Infinity; from the Circle of Divinity to the Circle of Mind; from the Circle of the Heart to the Circle of the Human; from the Circle of the Devic Multitudes to the Circle of the Mineral and Vegetable Kingdoms, and inclusive of the Animal Kingdom; and all said circles of life, upholding and interpenetrating all of the multivariegated realms of manifest experimentation, and becoming the Great Creative Circle of the Evolutionary ONE, emptying into Pure Emptiness.

25. OM SRI SRI GANESHA NAMAHA

    Om, salutations and reverences to thee, O Sweet Lord, GANESHA.

On this September day of the pot-bellied full moon
In celebration of "Ganesh Chaturthi"

__ __ ____
D   M   Y

October Full Moon

Durga

Holy Durga, personification of the Divine Mother, is certainly the most complex and powerful female divinity in the pantheon of Hindu gods. She is particularly venerated during the Navaratri (9-day) autumn festival, usually in October. Her forms are multiple. She is the manifestation of the Creative Force and the Power of Protection, and she represents as well the Shiva Power of Destruction, through which the High Ascetic Lord expresses Himself.

Divine Devi Durga's Battle for World Balance

Rays 2
3
1

Great Goddess Durga and Her Golden Simha

Devi Durga Puja
October Autumnal Full Moon

Part 1

Birth of the Devi

I
Creepy Aeval

In the long luminous fortnight of the primeval eternity of cosmic aeons ago
The good Gods, feeling that something was amiss, looked protractedly about,

And discovered to their absolute amazement that some hoary creepy aeval

Was slipping stealthily out of some concocted cracks, contrived into the occult configuration of time,

And that some offensive suspicious slime was starting to ooze through slightly,
And to contaminate irreverently the sacred corridors of free aeternal Space.

II
Mahishasura Attacks

And so it was, soon after, that Mahisha, the Buffalo-King of the dreaded and dreadful, demon Asuras

Burst through the ramparts of Heaven with his horrible horns and demoniacal roar

And with his hellish hordes of heinous dark spirits in tow, engaged Holy Swarga in terrible battle,

And so it was that one after another of the unready battalions of Heaven fell deplorably under the horrible onslaught.

And the good Gods were feloniously driven out of their legitimate abodes,
And deplorably depleted and totally humbled, were extradited and balefully banned out of the now mournful eternity,

And were shamefully sent to live as servers of time and mere mortal men upon the lowly earth...

Indeed, an unbearable demotion and an abject deposition, a base mortification and a terrible humiliation for the once proud, apodictic governing gods.

By the venerable right of unequivocal vanquishment, Mahisha, the terrifying and blustering Buffalo-King, now enthroned himself as the Divine Dictator and Supreme Sovereign of Swarga,

Thereby, dishonorably displacing Lord Indra, and becoming the infamous successor of the former Incontestable Ruler and Apical King of Heaven.

III
The Gods Radiant with Rage

The profanation of the sacred space of Sempiternity by the swarthy swarms of unholy Asuras thoroughly affronted and infuriated Supreme Shiva, Valorous Vishnu and the Unblemished Brahma.

And their Godlike, golden faces glowed together and were radiant with rage,

And from the holy fire of their fraternized furor, a most powerful, totally indestructible 'Supreme Shakti' mushroomed forth as a consummate 'Pure Potency', deep-rooted and muted in the dark-blue spirit of their Sheer Silence and Unmitigated Emptiness.

IV
Giving Form to the Devi

And from out of the luciferous anger of the 33 good Gods of the devic world, a luminescent entity took a diaphanous female form, within the tangible darkness of Mahavishnu's dream of Non-Existence.

From out of Shiva's fiery energy fulminating forth, her effulgent face took a delicate and comely form.

From out of Vishnu's loving look, languishing upon her subtle silhouette, her shoulders and arms took a lovely sloping shape.

From out of Brahma's bright beam of scrutiny her fetching feet took a fair configuration.

From out of Chandra's lunar luminosity came the sublime creation of her beauteous breasts.

From out of Indra's heavenly irradiation kindled forth her winsome waist.

From out of Varuna's oceanic effulgence of water and ambrosial wine undulated the shine of her strong sensual thighs,

From out of the blessings of Bhumi's light came the earth's gift of well-proportioned, shapely hips and bewitching, becoming buttocks.

From out of Yama's masterful power over death and his glowing triumph over darkness, arose a sweet salutation to the powerful growth of her perfumed, lustrous hair.

From out of Surya's solar blazing came the beautiful bright which were to become every one of her perfect twinkling toes,

And from out of divine benignity he further bestowed upon her skin the spreading golden rays of his own radiant brightness and dayshine transparency.

From Kuvera, Lord of Quickening Wealth, came the classical Aryan nose which inspirited and intensified the light of her awesome beauty and popular appeal.

From out of Agni's incandescent core came the bright solar luminaries which took the bewitching form of her twin fiery eyes, and her all-seeing luciferous third.

From out of valiant Vayu's wind of 'Vach' came the exquisite formation of her ears, and with it the gift of Ganesha's universal audience.

And from passionate Prajapati, the colorful co-creator, came the gleaming white teeth which bestowed upon her a divinely-sweet, pure smile and instilled in her the confidence of a sure victory every time.

JAYA! JAYA! JAYA!

V
Arming the Devi

And since it was the Divine Devi's destiny to engage in endless and untiring battle with the involutionary forces of inequity and evil, and to become the hallowed, Invincible Warrioress,

The good Gods got together once again in an occult conclave and decided to issue her formidable weapons in order to equip her warrior hands with a solid, sure defense.

And from his primeval Trident, Shiva fashioned a deathless and death-giving trident, and gave it to her,

From his primeval Discus, Vishnu contrived a razor-sharp, fatally-cutting, gyrating cakra of death and destruction, and gave it to her.

From his original Water Pot, Brahma patterned an absolutely beautiful and impeccable water vase for the execution of her daily ablutions and purification practices, and he also gave her a string of transparently shimmering, crimson colored beads.

From his primeval Flaming Spear, Agni, carved and forged a super-penetrative, indestructible, lethal javelin of fiery death, and gave it to her.

From his primal Lightning Vajra, Indra, formed a formidable, undefeatable thunder-hammer, and gave it to her.

From his hoary Staff of Fate and Finality, Yama cast a staff, (kala-danda), of sure destiny and inevitable finality, thereby spelling the death knell of evil, and gave it to her.

From his ancient Bow and primeval Arrows in a quintessential Quiver, Vayu, (Maruta), formed a bright bow and shaped innumerable arrows of light to rain death and destruction upon the dark enemy, and gave it to her.

From his Noachian natural Noose, or Lasso of Light, or Auspicious Pasham, Ambupati, (Sovereign of the Waters), contrived an inescapable, effulgent noose to capture and reign in all evil and with which to strangle the unholy, and gave it to her, (in a red quiver fit for a queen).

The Science of Full Moon Invocations

From his archaic Conch, Krishna freshly fossilized and formed a beautifully-sounding calling conch for the summoning and rallying of her warriors to decisive victory.

From his ancient Axe, Vishvakarma concocted a lightweight, razor-sharp, light-shimmering, unbreakable axe of lethiferous intent, and gave it to her.

From his time-honored, legendary Sleek Sword and time-tested Shineful Shield, Kala forged an unvanquishable, immortal, dazzling sword, and a divinely-favored and blessed shield of blinding gleam, and with a mysterious secret smile, gave it to her.

From his eolithic Well of Wine, Kuvera fashioned a perdurable, ineffably exquisite cup, full of ambrosial wine for the purposes of immortal youth, enchantment and bewitchment, as well as spell-binding magic, and gave it to her.

From his endless Oceans, Varuna arranged for the Goddess wonderful garlands of supernal lotuses to wear as a timeless reminder of her absolute purity of heart, and gave these to her.

From his hoary Himalayan Habitat, Himavat, or Shesha, the high-mountain Snake-God, searched far and wide for the strongest and most lordly of snow lions, and then commanded him to serve loyally and ferociously the great Goddess Durga, and he gave the noble beast to her... as a most potent vehicle for her to ride upon.

And it seemed that all the Gods in heaven had cooperated in accommodating the Goddess Durga with an astounding array of sublime weapons; and her Divine Female Form, bedecked with bedazzling jewels from the sacred, churning Ocean of Milk shone with a supernal shining... as she brandished forth her weapons, strutted her stuff, and roared her victory call... and her thunderous laughter and great, booming voice filled all the heavens and made the three worlds of creation convulse and shudder with wonder.

Devi Durga was ready to do battle and to bestow death upon the hordes of evil which had swarmed over Swarga.

Jaya! Jaya! Jaya! 'Victory to thee, O Goddess', the divine devas broadcasted forth their joy, as their ecstatic cries rended the ethers, and the wise rishis bowed their heads in deep musings and were filled with heavenly hope, as Devi Durga walked dauntlessly by.

VI

Mahadevi

And so it was that out of the expressed energy cognized as Creation, or the manifested veil of Mahamaya, came the Devi-Kali Goddess.

She in turn, by divine invocation called upon the combined will and creative forces of the Gods, and then she transmogrified into the 'Divine Durga', that is, 'pure, raw, manifest power', the concrete perceptible energy of mother earth, now immortally matter-ing for the sake of mortal, evolutionary man and the good of all beings.

Part 2

The Devi at War with the Demons

I
The Devi's Walk

As the three worlds shook from her spontaneous, stentorian laughter

And the earth itself quivered on its axis from her giant resounding footfalls,

Mahishasura woke up, startled by the titanic tumult and the eerie atmospheric agitation

And he bellowed forth a reboant, booming, bitching bellow and a rank, crashing curse.

"Who dares disturb the calm repose of the King of Universes and Supreme Lord of the devic dimensions?

Who dares create this ubiquitous uproar and cause a cornucopia of chaos in my extended, royal kingdom!?"

And tempestuously upset and brawlingly angered he stormed out of his palace to pounce upon the source of the escalading surge of sound and fury that was racking both sky and sea, and abjectly mortifying the earth.

And there, undisturbed, stood the dazzling Devi, straddling the three worlds with the scintillating shine of her golden omnipresence,

And permeating all with the harrowing power of her splendorous 'Shakti'.

Resplendent and intimidating she was, as her God-consecrated crown scoured the sky,

And powerful was her indomitable stride as it pummeled, smashed, and shook the perturbed and stunned, planet earth.

II

Devi and Mahishasura Engage in Battle

Face to face, with a clashing confrontation and a crashing clatter, mortal combat was engaged in by the Devi

And Mahishasura, with his millions of minions of every sort of maddened, murderous soldiers, shielding him.

Elephants, horses, and chariots circled around the Devi as scores of activated deadly weapons,

Such as swords, axes, clubs, spears, bows and arrows, and what not, tried to do her in.

The battle was deafening but the Devi sat strangely silent and very still on her regal lion,

Screened and protected by an unusual, magical circle of fiery refulgent light.

None could attain her as she dashingly bestrode her lion, and lashed out right and left with an uncommon calm and abandoned fury,

And she truculently thrust a resolute path through the thick of the battlefield, as if it was made of paltry butter.

III

The Battle Lengthens

Like a coruscating fire blazing crisply and easily through grassfields,
She fought and hacked her way through, with limbs and heads, by the thousands, dropping by the wayside,

As rivers of blood, knee-deep, piled up in gory pools, and made their way pathetically, as sorry red rivers, through a completely ravaged countryside,

Which was presently heaped mile-high, with the awful-smelling, corrupt corpses of the unholy wicked.

Devi made holy mantra on her bright mala as she banged upon her war drum,

And blew her unnerving and ear-benumbing victory sound, into Krishna's mind-stupefying killer conch.

And with each relaxed breath she took, in that short mortal, resplendent respite,

Would thousands upon thousands of demon battalions, many limbless and headless, rise up again as the fiendish, fighting undead.

And the Devi would rain endless clouds of deadly arrows and shower countless scores of pernicious spears upon them,

Till the very air whistled a savage death song, and the very horizon became dark and aphotic with the sheer number of relentless, lethal projectiles.

IV

And the Clash and Clamor Continue

Devi Durga, the divine destroyer was dispersing, debacling and drubbing Mahishasura's mighty army to shreds, and into shameful military impotency.

His best generals, even the awe-inspiring bowman commandant, Chiksura,
Who had twanged a veritable Himalayan avalanche of shrill ringing arrows, faster than the eye could see toward her nubile form,

Had all been cut down adroitly, like the odious dogs they were.

Commandant Chiksura, after having all his arrows halved before they attained their mark, and having his chic chariot undercut right from under him,
Ran forward on foot like a charging berserker, and thwacked mighty Simha on his head,

And with a mighty protective roar, the wild Devi raised her thunder vajra, and smote Chiksura's skull into an array of bloody, flying bits of bone and brain.

And then Mahishasura, beyond himself with red rage and murderous impetuosity,
Took the fearsome buffalo form, and sprung into a veritable whirlwind of whirling death,

Stamping and trampling the warrior gods under his hellish hooves, and throwing thousands into a gory and smashing death, with his Mephistophelian horns…
Even to gouging out entire mountains, as if they were mere morsels of chunky peanut butter clumps and rice-krispie balls,

And he pummeled and battered the planet earth to pulverized dust with his colossal, beelzebub hooves,

And the sky itself was badly cut and the ethers bled, and innocent clouds fell to earth like badly wounded, abandoned fluffs of non-spirited emptiness.

And Mahishasura's endlessly undulating, terrifying tail reached out towards the oceans and seas,

And whipped up one gigantic wave after another, until great cataclysmic tsunamis ruthlessly battered the earth's unsullied shores.

And his fulminating breath in red-hot blasts, put to fire the forests of earth, and pitilessly razed them down to mere blackened toothpicks,

And with his wrathful breath from Hell, he blew over entire hills and hillocks, as if they were made of mere piddling snuff.

And a pandemonium of evil portent, and an atmosphere of indescribable chaos reigned upon earth,

And the whole human scene thickened with white fear, as an overt oppression and utter hopelessness, settled dismally and darkly over the consciousness of man.

V

The Final Outcome: Mahishasura's Death

And the Devi bellowed out an ear-splitting and sky-sundering, unearthly sound of "Enough!"

Mahishasura turned, espied her, saw red, and rushed upon her in a beastly blitz of blind, bristling rage.

And within a microsecond of dynamic meditation, Devi Durga swiftly invoked all of her aggregated superpowers, and arrogated to herself her most formidable and horrific form.

And now raged a most mad and totally terrible, exhaustive battle between the chosen representative of the dark forces, and the divinely-endowed and godly-empowered champion, of the forces of Light.

So redoubtable, distressing, and monstrous was the contention between them, that battle reverberations shook the three worlds like trembling leaves on a tree,

And the devas, uneasy and disquieted, shut their eyes in frayed dismay, and directed their prayers to the supplication of the deathless gods for support.

And the redoubtable Devi with lightning speed, suddenly strung out her auspicious pasham, and lassoed the becrazed God-bull, and she tightly tied him up quicker than any rodeo champ ever could.

But no sooner than a fraction of a second, lassoed and bound-up, Mahishasura split from the habitual bull form,

And slipped into the spitting, growling, snarling, and scratching muscular body of an ogreish, titan tiger.

But before he could even swing a powerful deadly paw, Devi Durga pressed her shiny, sleek sword into his thick neck, and severed it imperviously on the spot.

But quickly arising from out of the absent neck's black gaping hole, Mahishasura shot out like a cannon ball and took the shape of a mighty warlock warrior, and brandished threateningly in hand, a barbaric gleaming saber,

And with her eyes bloody red and scowling with fervid fury, the Devi, having had more than enough of this titanic tussle of godly sorts between good and bad,

Shot him through heartlessly with a thousand body-piercing fatal arrows, and immediately, the paladin knave of darkness, fell decidedly dead.

But from out of the expired warlock-warrior form, a giant and grotesque, furious elephant of demonic configuration took shape,

And Devi, even more outraged, mercilessly chopped off his trump of a trunk, and turned it into a humiliated trumpet,

And roaring with rage at this new insult and with his pride on the line, Mahishasura reassumed his original King-bull form.

But Devi Durga was already ready, and before he could pull together the full force of this first, yet final form,

She leapt effortless to the very vault of the sky, and came back down swift as lightning unto Mahishasura's big burly neck,

And she pressed down her fair foot with indescribable strength and immovable gravity upon it,

And with a complete assurance, she tilted back her proud head, and roared a great rolling laughter unto heaven's hearing.

And as the bull King-demon squirmed right and left and desperately struggled to free himself,

Durga pressed down even harder and increased the pressure until he howled in unearthly pain,

She then hoisted high her celestially-forged splendorous sword and brought it down indifferently in one implacable master stroke upon his captive neck.

Mahishasura was dead.

VI

A Cheerful Celebration

And the heavenly hosts of devas extolled and exalted the Devi;
And the rishis rejoiced and the sages sighed and smiled;
And the gandharvas chanted sacred mantras and they played music and sang in gladness,

And devi and deva danced with delight in cheerful celebration of the Goddess Mahadevi's valiant victory.

> Jaya, Jaya Jaya!
> 'Victory to thee, O Great Devi!'

VII

Homage to the Goddess

Without delay, all the good gods and divine devas hurrahed their joy from on High,
And everywhere on happy earth was heralded the glory and greatness of Maha Durga.

Delighted devas came each in their own turn, to give homage to the victorious Goddess and to sing joyful hymns of praise and glowing songs of gratitude.

They garlanded her with felicitous recognition, and adorned her with glorious, glowing flowers from the perfumed garden of celestial Nandan.

And with hearts full of humility, they sumptuously anointed her with sacrosanct unguents, and honored her with splendorously sweet, sanctified incense.

Part 3

Invocatory Conjurations to Mahadevi Durga

I
Mahakali, the Black Battle-Goddess

'O Divine Devi, thou who hosts as thy holy garb the three divine manifestations of the One Absolute Power, known as **Maha Maheswari**.

Thou, who art the Great Emptiness filling the illimitable spaces of the universe...

And thou, as Mother Mahakali, who art the bright, Black Goddess depicting the primordial darkness,

And thou, O Goddess, who symbolizes the primal gestation and who is regarded as an original element of creation, art amassed in terrible potential form, as the tamasic, parturient aspect of the Dark Devi.

Swarthy she, of ten faces and ten feet, of black raven and sable skin, and atramentously beautified with becoming jewels and armed to the teeth with vanquishing weapons.

Swarthy she, who dearly loved Mahavishnu and induced him to rest and who, later on, would be moved to holy action by Brahma and the supplicating gods, and sworn to destroy all unholy forms,

Such as demon despots, demoniac embodiments, and other contemptible manifestations of evil.

Swarthy she, who is the universal 'shakti' and the mother of illusion, the magician of creation,

She, who guards and protects the One Truth and occults Reality away from those who are gross of mind and impure of heart,

She, who is the primal personification of Maya, and who is the hidden, nubiferous power of Vishnu;

She, who is respectfully and fearfully known as Mahamayi, and she, who is sole mistress of the Great Veil covering the all of creation, known as Mahamaya.

Unless Goddess Mahakali is appeased and pleased, and can be persuaded by devotion and puja to part with the Lord,

The Light which is the Lord within us, will not raise its fiery head, and blaze forth to annihilate the mean machinations of the ego,

And subjugates the enemy which proclaims itself as seminal ignorance, or primeval evil.

O Mahakali of ebony emptiness, first great aspect of Mother Maheshwari, grant us thy blessing, and guide us unerringly to Basic Goodness,

And may thy watchful eye, and ready hands, protect us always from the wickedness,

Malevolence, and baneful iniquity of all sinful separativity and duping deviltry.

II

Mahalaksmi, the Coral Red Warrioress

O great, second aspect of the Divine Devi, thou who art coral red in color, and who displays either nine, or eighteen arms and hands, and who wields a weapon in each, as a gift and boon from the gods.

Born were thou from the fiery righteous fury and synergistic powers of the great deva gods combined.

Thou, who has the fearless will to fight the dark forces to the last death blow, and final, vile, expired breath.

Thou, who willingly manifests the heroic 'rajasic' aspect, or dynamic action principle of Mother Maheswari,

Thou, who art recognized as the generous giver of life, and the supreme savioress of all who are sorely beset by unholy hosts,

Thou, O Great Goddess, who sports the revered appellation of Mahalakshmi, or Divine Devi.

Sanguine she, who like Athena, is the terrible courtesan of war, and the supernal white warlock,

The ultimate warrioress of battle waged, and who spreads the red cast of blood everywhere into the dusky horizon of holy contention.

Sanguine she, who ferociously felled the fearful and nefarious bull-King, Mahishasura, the savage manifestation of utter narcissistic ruthlessness,

And he, who was the immoral, fiendish vehicle of the jungle dictum that 'might is right', no matter what.

Mahishasura, who's primeval, brute-force morality was the base principle
He inflicted with lacerating impassion upon all personalities, whether they were gods, or mere mortals.

'No opposition was his composition.' The small selfish self lorded it meanly over the Noble Self.

So awesome was his great strength that he superciliously struck at Swarga,
And divided the gods, and dispersed them successfully; and he separatively weakened them, all over the four corners of the battle-wasted dimensions.

However, Mahishasura met his match as well as his waterloo, with the synergistic power of the pantheon of thirty-three gods,

Acting as One within the implacably Divine Will and destructive Dynamic Action of the Devi Mahalaksmi,

Who, after the combat smoke had died down, and with Mahishasura destroyed,
Became known and thereafter honored, on earth as well as in heaven, as the Great Savioress, Mahishasuramardini.

The brutish bull-King, Mahishasura, as the symbol of embodied evil, stood metaphorically for primal ignorance and blind unseeing darkness.

He stood for the blight of separativity, or the sin of separateness, as well as raw, unbounded egoism,

And the mindless (cruel) ambition for power, and the myopic greed for possessions.

O, Goddess Mahishasuramardini, help us, dear Divine One, to conquer these negative states within us,
And to master our mean, small self, metaphorically personified in the demon-King, Mahishasura.

III
Mahasaraswati

Divinely intercede for us, and aid us in gathering together all the good, and godly forces within,

And to apply them with lighted will and noble want, to the ultimate destruction of all darkness and selfishness, ignorance and ambition

Which might blindly fester as spiritual sores within the ramparts of our aspiring being.

O Divine Devi Mahaswari, manifesting now as the silvery, shining, and third aspect of the Supreme Goddess,

Holding in her eight enlightened hands, first, the illumined sphere which enounces from its fiery core the eternal Aum,

And in the second, the transpiercing trident of Truth Eternal, and in the third,
The ploughshare which furrows out the righteous path for the true sadhaka,
And in the fourth, the conch which rallies forth the inspiring company of fellow spiritual warriors,

And in the fifth, the discus, the divine sudarsana which sharply slices through all that is unreal and sets the stage for the Real to appear,

And in the sixth and seventh, the bow and arrow which unerringly strikes to the bull's eye of the pure Heart,

And in the eighth, the pestle, the pulveriser of evil, the grounder of negativity, and the pounder of the selfish ego.

All of these positive, golden qualities herald the coming of the third feminine principle of Maheswari,

Which is that of the supernal Goddess Mahasarasvati, who is as luminously resplendent as the mysterious shine of the full autumn moon of October.

Auriferous she, who is undisputedly, the paramount paragon of physical perfection,
And who is, unquestionably, the ultimate quintessence of divine beauty,
And who is, incontestably, the uttermost enigma of womanly wiles,
And who is, incontrovertibly, the mature pearl of feminine expression.

Auriferous she, who is the building power of intelligent, creative work, and the primal inspiration and inspiriting potency of all arts, music and creativity.

Auriferous she, who is the enlightened embodiment of divine order, synthesis, and cooperative organization,

And who is the celestial evoker of that unlimited strength to be gotten through a coactive unity, and the forming a harmonious, united front.

Extensive are the prompt, positive, refulgent exploits of her great, heroical heart.

Bless us, O Mahasaraswati, with thy unsurpassed beauty, thy inspiriting intelligence, thy igneous inspiration, thy purity of Heart, and thy love of Truth.

We honor thee for thy Divine Impeccability, and we bow to thee in humble deference, and render to thee our esteemed salutations.

IV

Devi Durga
In Endless Engagement Against Evil

With great bravado and brouah and lots of blood, brain, bidding, and hellbent obduracy, and leaving behind a trail of beastly bewailment,

Auspicious she, of power supreme and mandated by heaven, endlessly engages in battle with evil,

And who has, over the ages, pummeled, punished, and destroyed a host of unearthly demons...

And notably to be found amongst them, are the fiendish likes of Nisumbha and Sumbha, Munda and Canda, Raktabija and Dhumralocana... all maleficent, demoniac, utterly narcissistic, asinine Asuras,

All acrimonious archetypes of crystallized egocentric entities who are entirely self-serving, self absorbed, self-admiring, and self-satisfied... and hopelessly addicted to the glory of arrogant power and the principle of conceited, (often cruel), play and pleasure.

All the stages of primal evolution, from gross to subtle, are metaphorically represented and portrayed,

Through the prideful speech, overbearing emotions, vainglorious actions, and arrogant mental states of these slimy, haughty Asuras.

Dhumralocana

He was dumb and cloddishly daring, and with an sizable army of 60,000 non-thinking asuras, this smoky-eyed, brute strength monster, and kowtowing messenger of Sumbha and Nisumbha,

Had been sent approbriously by these two latter generals to fetch the dazzling Devi by force, and bring her back, ingloriously, and shamefaced, by her golden hair.

The Devi decided to put an abrupt end to this pig's big, boastful, bragging conceit, and Dhumralocana, who dastardly depicts base ignorance and buttoned-to-the-top, obvious egoism,

Was destroyed by a 'humkara' a single, scornful sigh of 'Hum' from her pure Heart!

A fierce frown derisively furrowed the Devi's burning brow, as Dhumralocana withered, and was no more… and only a heap of unholy ash meanly marked the impious spot where it all happened.

Canda and Munda

Consequently, the cruel chiefs, Canda and Munda, mean beasts of most malicious and deadly intent,

Were sent by the crazed and enraged Sumbha and Nisumbha, after hearing about the stupid death of Dhumralocana.

They came at her with an awesome array of deadly arrows and shining swords, and the clash and clamor was most dreadful.

And it was then that the Devi's bewitching smile suddenly went dark as did her entire skin, which looked shinely singed,

And posthaste, a hideous, horrendous Kali raged from out of the Devi's fiery, furrowed forehead,

And straightaway she took them in a tremendous deathhold and dashed them dead against her divinely adamantine bare black breast,

And then throwing them scornfully in the air, as if they were mere barbaric babes, she caught them by the hair in the downfall,

And in one single stroke, cut off their astonished, open-mouthed, ghastly heads.

From that triumphant moment on, the dark goddess on her golden simha, was called Chamunda, the Fearless One.

Raktabija

The perpetually self-regenerating Raktabija Asura, saw his armies ransacked and decimated by the Devi,

And it was with absolute assurance that he strode rashly and unhesitatingly into the bloody fray,

Knowing that for each drop of blood that plunked down to the planetary soil from his powerful body,

An unholy Asura of equal and formidable stature would resurrect on that very spot.

The Devi smiled an inner smile, and she laughed uproariously out loud, as she immediately sized up the situation,

And calling on total-death Kali once more, she made this suggestion to her:

"O chomping Chamunda, benign vampire of the gods, stretch out your infinitely-wide mouth and with relish, regale yourself upon the blood which flows from this bedamned Asura.

Open wide and drink up, my dark beauty, and let not a drop of blood drop to the ground unswallowed by thee!"

And this she did, and absolutely obeying the Devi, she drank and drank and drank and narry a drop, not one, fell to earth.

And with his root power now neutralized, his seemingly endless strength presently left him,

And the Goddess Devi promptly afflicted his body with a thousand arrows of light, and five hundred fiery darts, and one hundred flaming spears.

And he died, all crumpled up, his impotent corpus completely caked with his own dried-up blood.

Sumbha and Nisumbha

Sumbha was supernaturally enraged, and Nisumbha, completely numbed, but on fire inside, in seeing their armies so skillfully slaughtered,

And their best warriors being bested, and forced to take dastardly flight.

Thunderbolts flagellated the ground with fulminating fires and thunderclaps roared and ripped through the air,

As the battlefield filled with the chaotic cries of war and the pitiful whimperings of the wounded.

The divinely infuriated Devi decided to spice it up, and cut Nisumbha into burnt slices of smoke meat,

As he dared tempt fate and tried to harm her valiant Simha, her loyal and royal snow-lion, and her indefatigable vahana.

And then she slowly turned to the sinuous, imperial Sumbha and surprised him with a stupendous, thunderous clap of her expanded, open palms,

Which categorically shook the screaming sky right to its upside down roots,

And with a whooping war-cry of fated victory, she rendered asunder all deadly arrows aimed at her by the grieving and enraged Sumbha,

Who had seen his brother shabbily slain and his whole army execrably slaughtered.

And so without constraint, was his crazed and uncontrolled anger, that all the friendly goddesses feared for the Devi,

And they bravely bounded by her side to protect and gird her in her noble defence.

And Sumbha, contemptuously called her out as being craven and cowardly, for fighting with a friendly multitude of warrior goddesses by her side.

And turning to face him directly, the Devi became deathly still and smiled, and gave him the following reply:

"All alone am I, O vile and wicked one, and have always been so, O defiant defiler of all that is Divine.

October Full Moon

These brave and superb warrioresses who stand by me and defend my person, are but my own supernal Shakti's expression eager for battle and sure of victory.

See for yourself now, as they calmly reenter and are reabsorbed by my Blessed Form.

Their supercharged energies are, even as I speak, being quietly consumed and gently assimilated by me, the Mahadevi, the Goddess Durga, standing alone before thee, battle-ready and eager for thy blood.

The fighting was terribly fierce and the dust of earth gyrated about and rose like giant whirlwinds filling the skies with demonic froth and divine fury.

And at last, upon a supernatural surge of stupendous strength, the Great Devi lifted him up high into the Immensity and cast him down again,

As he hurled his disarroi, (like a betrayed coyote), at the full moon!

The earth stood momentarily stunned, as the whole planet trembled and rocked on its axis.

And within this opportune frozen instant, she sleekly sank her sacrosanct spear into his blackened heart,

And he screamed an awful screamless scream, and then lay back, mouth agape, totally lifeless.

And a clarity immediately returned to man's consciousness and an illumined calm overtook his mind,

And an OM'd Shanti suffused humanity's heart, as the worlds upon worlds were once more made safe, and evil was sealed down for many enlightened ages to come.

V

Goddess Durga Invocation

'O Mahadevi, O Great Goddess Durga, we sing praises to thee and revere thee, and unceasingly repeat thy Divine Names.

Please be pleased, as we do devotional puja to thee and worship thee with a humble tumble of devout prayers.

We ask for the strength of thy protection always, and beg for inner guidance upon the spiritual path, (which, to all appearances, seems to be the one we are now following).

May thou deliver us from all fears, and keep us safe from harmful foes.

May our life be imbibed with light, and may we invoke the light and walk out of the darkness into the truth of day,

Where the one reality of the conquering Lord abides, and all the transgressions of ignorance, and the sins of separateness,

And the bad doings of the world, and all the evil work of the wicked have been rendered null and void,

And have been compassionately compensated for, by Heaven's beneficent grace.'

'O Devi Durga, bless us, bless me, as all the good gods in cooperative synergy have done unto Thee,

And both heaven and earth have been the better for it, and forthwith and forever, all will be well for the residents of all worlds, even in the probable worlds and parallel dimensions.'

VI

Sri Durga Mantra

Mantra 1   "Gauré Narayani namoh stu te namah."
"Three-Eyed Narayani, I (we) salute and bow to thee."

Mantra 2   "Durga Kshama namoh stu te namah."
"Oh, Durga, manifestation of forgiveness, I, (we) bow to thee."

Mantra 3   "Jaya Jagatambé, Gauré Durga namoh namah."
"Hail to thee, Mother of the world, O Golden Durga,
I (we) salute thee and bow to thee."

Mantra 4   "Kalee Durge, namoh namah."
"Palana Karanee bhava bhaya haranee."
"Namastasyai, Namastasyai, Namastasyai."

"(Black) Kali Durga, I (we) salute and bow to thee."
"You shelter me and save me from fear and death."
"I (we), bow to thee over, and over, and over again."

Mantra 5   "Namo Durgaye Shree Kali Ma."
"Salutations to thee, Oh Durga, who is also Sri Kali Ma."

Mantra 6   "Ya Devi Sarvabhutesha, Shantirupina, Samsthita."
"Salutations to the Divine Goddess (Mother), who abides in all
beings in the (sweet) form of peace."

Mantra 7   "Durgati Naashini Durga, Jaya, Jaya!"
"Hail and Victory to thee, Oh Durga."

The Science of Full Moon Invocations

Mantra 8   "Jaya Jagatambe Seetah, Radhe, Gauri Durge, Namah, Namah."
"Paalana Karani Bhava Bhava Haranti
  Kalee Durge, Namah, Nameh."

"Hail and Victory to the Mother of the world, also called Sita,
  Radha, and the (Golden) Durga."
"I (we) bow to thee again and again... to the Goddess
  who lifts me (us) up and saves me (us) from the fear of this world.
  Kali Durga, I (we) bow to thee over and over again."

Mantra 9   "OM Sri Maha Sarasvatyai Namah."
"OM and Salutations to thee, Oh great, (and beautiful),
  Goddess Saraswati."

Mantra 10   "OM Shree Matre Namaha."
"Jai Ambe Gauri."
"Jaya Shyaamaa Gauri."
"Jaya Jagatambe, Hey Ma Durga."

"Salutations to the Holy Mother."
"Salutations to the Golden-Skinned Mother."
"Salutations to the Ebony-Black Mother."
"Salutations and Victory to Ma Durga, the Mother of the World."

Mantra 11   "OM Naraayani OM.
  Jaya Jaya Jagatambe Maa Durga.
  Hé Maa Durga.
  Namastasyai, Namastasyai, Namastasyai."

"OM, Three-Eyed Narayani.
  Hail to thee and Victory to thee, Oh Ma Durga, Mother of the world.
  Oh, Holy Mother Durga.
  I (we) render homage to thee over, and over, and over again."

October Full Moon

Mantra 12   "Jagaan Maatar Maata, He Bhavani, namoh, namah."
"Oh Bhawani, Oh Ma, (Supernal) Mother of the World,
  I (we) salute and honor thee!"

Mantra 13   "Ya Devi Sarva Bhuteshu Vishnu Mayeti Shabditaa"
"Namastasyai, Namastasyai, Namastasyai"

"The Goddess who pervades all things exists also in the form
  of the Lord (Vishnu)."
"I (we) bow to thee again, and again, and again."

Mantra 14   "OM Lakshmi lajje mahaavidhye
Medhe Sarasvati wari."

"OM Lakshmi, Who honors Sarasvati,
Upholder of the Power of Speech."

Mantra 15   "Swarg parwargarde Devi
Narayaani namo stuste."

"Devi Narayaani, who eliminates darkness
and transmits the Light, I (we) salute you."

Mantra 16   "OM Dhum Durga Jay Namaha."
"OM and Salutations to thee, Oh Durga."

November Full Moon

Kwan Shih Yin

In China, Kwan Yin is the feminine counterpart of Avalokiteshvara, a Bodhisattva embodying the essence of Compassion and the Infinite Love of all realized Buddhas combined. She has taken to serious task the Bodhisattva Vow to liberate all sentient beings from samsaric suffering.

Kwan Shih Yin's Initiatory Impulse of Always Benign, or Compassionate, Action

Rays 1
3
2

Kwan Yin,
Inexorable Mother of Mercy

Kwan Shih Yin
November Autumnal Full Moon

Section I

Part 1
Kwan Yin, Mother of Mercy

The Science of Full Moon Invocations

- 1 -

O Kwan Yin, Goddess of Mercy and Mother of all who may be in real need

You, who are honored all year round and whose famous Brother's birthdate is said to be the 19th of March...

We, as an invocative group of conscious disciples choose to celebrate your Holy Name,

Now, (because of real need), under the auspicious influence of the full moon glow of the month of November.

- 2 -

You, who hears the cries of the world in its raw pain and unmitigated suffering, hear us now, O Goddess

We harken unto you and ask that you be amongst us in this spell of trial, test, and truly tough times,

The world is in need of your loving compassion and your living guidance in these times of turbulence,

Of incessant little wars, terrible terrorism, tsunamis, tornadoes, hurricanes, and all sorts of violent weather disturbances,

And awful earthquakes, and dreadful (predicted) viral pandemics, and epidemics of AIDS, and all sorts of sexually-transmitted, or other-related scourges,

And of appalling, pitiful poverty conditions, and of stark world-shameful starvation...

And all this because of world-disgracing selfishness, economic greed, and unbridled national ambitions,

And racial pride and prejudices, and governmental, tyrannical dominance,

And the twisted, corporate world of close-fisted competition, for the sake of the mere carrion of personal gain and of penurious profit,

And because of the intentional, frightful imposition of international political power over the rest of the shamefully-separate, One World.

Is Humankind destined to become the irreparable enemy of Mankind, the arrogant, unfeeling killer of planetary life and the inhumane destroyer of the planet?

- 3 -

O Holy Mother, you who are the equivalent of Matre Maheshwari and the Blessed Mother Mary,

You who are Kwannon's alter image and equal in compassion to your Bodhisattva Brother, Avalokiteshwara.

Hear us, O Mother of Care and of Caring for the world, and for all of its epochal, samsara-suffering inhabitants.

May your unequaled compassion bring light and relief to the burden of Mankind's planetary karma.

May your watchful eye protect us from the evil-eye of the dark, involutionary forces,

Which are ignobly intent on diverting us from the path of dharma and righteousness,

And are the cause of our ignorant mind crossing over to Mara, the Tempter's Camp,

Or, to heinous Shaitan's howling hovel of hell, and on to horrid insentient hellishness of hellspawn conduct.

- 4 -

Do cloak us, O Mother of Solicitous Concern, with your veil of unnodding vigilance

And protect us with your shield of purity, and please, hear our heartfelt prayer to you,

And salvage the confused world of present times from its samsaric woe and self-serving suffering.

May you, with the tender cast of your loving eyes, bestow upon us the thousand healing graces of your munificent blessing.

- 5 -

'OM Sri Matre Kwan Shih Yin Namoh Namah

Namastasyai, Namastasyai, Namastasyai'

'OM, Salutations to you, O Blessed and Beautiful Mother, Kwan Shih Yin,

We bow to you again, and again, and again, and we do honor you!'

Part II
Kwan Yin's Liberation

November Full Moon

- 1 -

As the majestic, Himalayan Maha-Avatar, BABAJI has felicitously freed and propitiously conferred upon the world

His occult female counterpart, who has since fondly become known to us as Avatara, Sri Mataji.

And the Lord Avalokiteshwara has also done likewise in liberating and blessing

His twin, (self-created), sweet lady consorts, the compassionate Green Tara and the Wisdom-born, White Tara...

So too are you, O noble Sister of Mercy, presently emancipated from being irredeemably bound to the Bodhisattva greatness of your prototype Brother,

That is, He who nobly bears the name of the Incomparable Kwannon, 'most compassionate Lord of utter enlightenment'.

- 2 -

The three 'San Ta Shih', or Sublime Bodhisattva-Beings, in male manifestation...

Who watch over both Humanity and the animal realm, and who display an obvious mastership over the forces of nature,

Who are the All-Wise... Lord Manjushri, the All-Beneficent, Lord Samantabhadra, and the All-Compassionate, Lord Avalokiteshwara.

And to the three 'San Ta Shih' counterparts, or Sublime Bodhisattva Beings in female form, who have basically the same supernatural powers, but needless to say, are expressed in slightly different modes and subtle nuances,

And who form today's titanic triad of feminine grace and mercy, and by whose names are respectfully known today as the Maha Mother TARA, Maha Mother MARY, and Maha Mother AVALOKITA, or the Heavenly Lady, KWAN (SHIH) YIN.

- 3 -

To the Divine Sovereign, AVALOKITESHWARA, rightly belongs the magnificent Potola Palace, as found in Lhasa, Tibet.

To the Divine Lady, KWAN SHIH YIN, or Padma-Pâni, that is, 'She who was born of the Pure Lotus',

Belongs the holy Pataloka Mountain, of the famed Buddhist pilgrimage island of P'u Tuo...

A sacred asylum, even unto recent times, of at least a thousand resident monks and over one hundred sanctuary temples.

Part III

Kwan Yin's Instantaneous Response
To Invocatory Pleas and
Their Immediate Benefits

November Full Moon

- 1 -

If ever a discomforted, suffering, scared, agonized, or guilt-ridden person calls upon the High-Born Kwan Shih Yin, from

The depths of his or her heart, she will clearly hear the cry for help, and hurry to his or her aid, and salvage that person

From his or her pain, ache, unease, or suffering, and this, with never ever a recriminatory stitch of imputation, inculpation, retortion, or blame.

Like nurturing Mother Mary, or the cherishing, Compassionate Tara, Kwan Yin's primary quality is a gentle, sweet maternal love

Tended towards all who call upon her Holy Name and ask for the comfort of her consoling Presence.

- 2 -

She is the 'universal gateway' to heavenly intervention and a sure shot for divine consolation.

The chanting of the mantras of Buddha Amitabha and that of the Lord Avalokiteshwara upon a mala, are everywhere heard,

But, Kwan (Shih) Yin, as the imminently approachable, Divine intermediary for all souls who are caught in (the pain) of samsara,

Is even still and ever yet, without the practice of constantly being chanted… everywhere revered, entreated and wholly loved, with not ever a trace of fear, nor even a timid trembling, in anyone's heart.

And she is ubiquitously adored and reverently worshipped on the familial altar of every simple, common household.

- 3 -

She is ever the benign facilitator of a person's spiritual insight and awareness,

And ever the beneficent, intuitive impetus and harmless catalyzer of Divine polarization within the human consciousness.

Immediate therefore, are the salutary benefits if you call upon her Holy Name, for a bracing boost

Of blest backing, or some fortunate sustenance in insufferable sorrow, or bleak intolerable pain.

- 4 -

She is the judicious, empathic, luminous Mother who magically transmogrifies

Into the form sympathetic to the person being helped, whether it be a child, a father, or mother,

A nun, a monk, a political personality, royalty, or just plain, poor folk...

She will always appear in the compatible prototypal aspect appropriate to the person being approached,

Who, in mind, is overwrought and overanxious, and ardently calling upon her for help.

So that, atmospherically, whomsoever calls feels in his person immediately safe, composed, calm and trusting,

And somehow, close enough to her, subjectively-speaking, to have complete confidence in her wise, (often telepathic), words of personal counsel and caring directives.

- 5 -

Similar to a child getting an almost instantaneous response upon summarily summoning mother,

The Divine Mother, Kwan Shih Yin, is famously known to be lustrously constant,

In the speed and fastness of her response to the tears of her human children,

And to the inchoate bewailments and laments of the innocuous animal world, crying for help and invoking healing ministration.

Part IV

Kwan Yin's Luminous Forms, Variegated Vahanas and Noble Ornaments

- 1 -

Tathagata tall and sempiternally slim, she is luminously clad in an ethereal robe of pure white, or Virgin Mary blue.

- 2 -

At the peak of her glorious crown she sports proudly, yet most humbly, an image of her Lord, the Amitabha Buddha… although at times, a tantric fire of pure Emptiness burns in the stead of His sacred Space.

- 3 -

In her graceful hands she may delicately hold a decorous 'willow branch', which lends itself to a certain metaphor symbolizing a natural bending with the wind of change, and leading to a necessary transformation of the personality.

It also lends itself to the ideal of not breaking under the illusionary weight, or maya of any adverse circumstances.

It is also recognized as a symbol of the pliancy and patience which are urgently required in the long process of a complete healing.

And finally, the willow branch is expediently utilized by her, for the aspersion and dispersion of the waters of life, which so preciously overflows her ambrosial vase.

- 4 -

In her graceful hands, she may carry, as well as cherish, a small, gracile, beautifully-attenuated vase,

Bearing some ambrosial nectar to be renascently poured upon the wounds of the world,

And which, furthermore, serves to symbolize the blessed waters that are to wash away the world's awful sins of insufferable separation and odious separativity,

And of which, her healing Amrita is destined to bring about a golden peace to Mankind, and a new homebred harmony to Humanity.

- 5 -

She is the archetypal, caring Female Harbinger heralding the imminent coming of a new, mythical Golden Age, wherein the Angelic Realm is to be brought into a closer contact and cooperation with an Awakened Humankind, within the enlightened, astrological kalpa of the Age of Aquarius.

- 6 -

In her graceful hands she may lightly clasp a *crystal* mala, (rosary), as her slender fingers slip from bead to bead, (personifying a peculiar attention paid to each and every person on earth), in a continual prayer for the relief, comforting, and soothing of the generic suffering of all beings.

- 7 -

In her graceful hands she may palm a beautiful, slightly gem-studded, Sacred Book, (or scroll of sacred Buddhist teachings), exemplifying the path of Dharma,

The central text, being without doubt, the Heart-Sutra Canon, which subtly divulges the esoteric pure source and divinely joyful wherefores of her compassionate, primordial origins.

- 8 -

In her graceful hands she may also, unostentiously, uphold a lovely lotus flower,

Symbolizing that she was unmistakably 'lotus-born' and that her spirit is one which is perpetually permeated by a 'padma-purity' of impeccably benign Intention.

And that, alike her Brother, the most compassionate Lord Avalokiteshwara, her mission has to do with the planting of Peace upon earth, and the implanting of Goodwill as the reigning rule of Intent amongst men,

But her 'modus operandi', being uncontestably tinged by the utterly-feminine and motherly-imbibed Bodhisattva penchant of dispensing freely the Buddha's blessings, and apportioning out her Heart's Essence of Compassion,

Which should fortunately, in resonance to Noble Avalokita, manifest bountifully to all, thereby helping to consummate her Bodhisattva Vow... of saving all sentient beings, till the very last soul is taken up into the Pure Land of Tushita Paradise.

- 9 -

She may be envisioned, graciously standing as a forbearant, feminine, all-seeing Bodhisattva Buddha, with or without, a thousand eyes and arms, in responsive similitude to Avalokiteshwara's seeing, sensing and soothing the pain of samsaric suffering, in all of the ten directions at once.

Or, she may be majestically sitting in the relaxed, royal pose on a feline Hou, (a mythical Tibetan-like lion), representing her consummate power and mastership over the laws of nature.

Or, she may be seen in the presence of another Venerable Vahana by her side, such as the Peerless Peacock, with its blue-green tail fully spread out, and the obviously stunning symbology of an All-Seeing Eye demarking each sublime feather.

Or, she may be soothingly carrying a gentle white dove, symbol of the innocent child and harmless soul, as well as being a pure metaphor for a desired path of peace and the generic way of harmony.

Or, she may be found to be masterfully riding upon a Sky-Dragon's powerful neck or mighty head, dramatically coming out of the clouds and proffering a silent basket of universal blessings and clemently offering, (especially to sailors), an assured protection from the waters.

Or alternately, she could be in the august company of an imperial Royal Dragon, thereby exemplifying an infinite sagacity and the magical power to transform into any form,

And in the midst of any action, or situation, always being acutely aware of everything, and all the while, consciously invoking a universal pathos for all of sentiency.

- 10 -

Her delectably delicate right hand may be shown facing slightly downwards, and palm outwards,

In the divine modality of a Bodhisattva bountifully blessing all those who may be in sincere need,

Or, on the other hand, the graceful gesture may simply represent the wish-fulfilling sign of some noble and needful request, being herewith granted.

- 11 -

She is always beautifully and becomingly bare of foot and is sometimes seen standing, (or sitting), on a pure padma, or lotus heart,

Betokening on an equal footing, a heavenly purity and an earthly humility,

And making of her mystically, an ethereal creature of the sky, as well as a bone fide Bodhisattva, compassionately touching the indigent planet earth.

- 12 -

She is often found, fondly surrounded by children, or seen tenderly holding, or devotedly nurturing a child,

And therewith, she has become the celestial boon and redeemer of barren women,

As well as the genteel goddess of fertility and ruthful deliverer of fecundity,

And as such, the epitome of true femininity, and the upholder of all fair women's undeniable rights,

And ultimately, the pure protector and gentle caretaker of all who are innocent, weak, afflicted, frail, sick, old, the very young, and naturally, all women who are pregnant of child.

- 13 -

She is at times, especially in her later bronze configurations, seen entirely naked,

With long limbs of infinite grace and sweetly curvaceous of form, and beautifully composed, and reposed in facial features,

And conveying to the world an active conciliatory compassion and a serene, sublime selflessness,

In open acceptance of the full responsibility of her Bodhisattva mission, and in her dynamically going forth to save all sentient beings from the desolation of pain, and the disheartening devastation of continual suffering.

- 14 -

She sometimes displays a bowl of white rice, or unshyly shows to all a sheaf of ripe rice,

As a candid symbol of her nurturing nature and her forthright promise of continual sustenance,

[And in noted afterthought, 'ripe rice', or 'corn kernels' in South America, is noted as being a metonymy for the designation of fecundity.]

- 15 -

Upon the temple altars she is often found flanked by a male and female devotee.

On her right side stands a shoeless, usually barefoot, young boy in the clasped-hands prayer mode position,

And who, through the flowing ages, has come to be known as the legendary 'Golden Boy',

Or, simply 'Shan-Ts'ai', the smooth-cheeked, bright-eyed lad of extraordinary means.

To her left, is devoutly positioned a young maiden, diffidently entwining her hands together inside her long, enormous sleeves,

And she, through the ages, has come to be known as the legendary 'Jade Virgin',

Or more transparently, as 'Lung-wang Nu', the downy-cheeked, flawless daughter of the 'Dragon-King'.

- 16 -

And at rare times, her ivory-colored cupped hands are seen as esoterically settled into the Yoni Mudra gesture,

Symbolizing, as a matter of course, the primordial gateway through which all beings naturally go, who wish to incarnate into this world, (or into any other world), alluding, undoubtedly, to the inescapable contingency of the all-pervading female principle in the Universe.

Although Kwan Shih Yin treats all beings with equity, She, as the Shining Defender of the (so-called) 'weaker sex',

Unreservedly throws her cloak of protection over literally all women… but most especially, for those women who openheartedly and unaffectedly call upon her for help.

And she unconditionally proffers to all who undeceptively inquire, a future bright pallet of alternative choices… whether it be a coveted career, a worldly life, or another of worldly renunciation, or of one featuring the common dream of a beloved husband and a desired number of children.

- 17 -

Similar is she to the Great Masters of the Heavenly Hierarchy in being Unique, yet so like unto Them,

In her being utterly enlightened in Mind, she stands yet, utterly humble in Heart.

She stands absolutely liberated from all possible haughtiness of Heaven, and holds herself erect, yet plenarily free from all passion and pride.

Her 'High View' clearly mirrored in consciousness, resonates a perfect emancipation and freedom from all mean sense of reprisal and base retribution.

Moreover, she tends to be highly sensitive and obstinately disinclined towards the utilization of disciplinary measures, and all of the so-called firm correctional methods, involving the use of strongarm techniques and 'tough teaching';

And this, even when appropriate punishment seems to be reasonably justified, on account of guilty indiscretionary deeds;

And even when severe lessons should be called for, as in the case of certain nefarious deeds, or iniquitous acts, and punishment is therefore looked upon by society, as being downright imperative.

She is, in her Divine Being, incomprehensibly and absolutely beyond the venom of vengeance and the vehemence of any sort of condemnation,

She simply loves, accepts the whole person, (no matter what), and spontaneously doles out whatever restorative comfort, cheerful consolation, and clear words of counsel, or wisdom, she can…

In order that a definite, gentle, and complete healing take place, and a turnabout beginning, or new benign direction, be amiably desired and righteously undertaken.

She is the divine, pulchritudinous, compassionate emancipator of prisoners, (even on death row),

And she actively responds to all sincere calls for better rebirth and reincarnational renewal made by all sinners…

If the penitent pleas which are fervently forged by them are compliantly delivered from the fiery altar of an honest repentance and an unfeigning open-heart.

Part V

Kwan Shih Yin
Sweet Savioress of Mankind

- 1 -

Resourceful and ingenuous is she in her infinite variety of dress, disguise, sex, shape and form, which

She appropriates to herself for the sake of somebody's salvation somewhere, everywhere, and anywhere… here or there, within the whole Hierarchical set-up of Creation…

On her well-travelled, planetary agenda of spiritual facilitation, Buddha-befriendment, and rescue mission of the Empty Bright… and of its beneficent revolution on evolutionary man.

- 2 -

She, who disapproves of religious dogma and the dogged ritual of narrow credo, and blind aspiration,

She, who compassionately enlightens the heart, enheartens the spirit, and emboldens the back,

She, who ardently buoys up the Divine Purpose and endears all beings to her Loving Presence,

She, who stirs up full spiritual arousal against inequity and injustice, and who spiritedly rallies together the species of humankind to selflessly serve Mankind.

She, who reveals her maternal affection to everyone, discloses her intimate face but to a few, imparts her immaculate heart but to the pure, and divulges her innermost secret self, only to the vast Emptiness.

- 3 -

O, omni-flowing heart of Compassion and lovely lighted lamp of Love,

O, tall white-robed Lady of the azure skies bestowing a thousand Graces.

O Blessed Lady, who has etherically kissed the empty breadth of Heaven, aloft the noble, august head of the silvery-moon's misty Dragon Breath.

O Blessed Lady, who is mystically and mythically said to be the Emanated Embodiment of the BUDDHA OF BOUNDLESS BRIGHT,

O Blue-White Imperatrix, who is beauteous and pleasing to the sight, who is high-principled and splendorous,

Who is the Paradisiacal Purveyor and Celestial Conveyor of Lord AMITABHA BUDDHA's great, capacious, coruscating core of 'Maha Karuna'.

The Science of Full Moon Invocations

- 4 -

O Illustrious Beauteous She, of the thousand eyes, and arms and hands, who is highly famed and universally honored,

As the ubiquitous, omnifarious Holy Mother looking in the ten directions at once.

O Sublime She, who supernaturally senses the afflictions of mankind and the woes of the world,

And who attends to the wounded hearts and tortured spirits, and broken bodies of all beings, with

The utmost of ministering tenderness and soft solace, and wise understanding and infinite compassion.

- 5 -

O Venerated One, who is said to be the High Mistress and Sure Skipper of the BUDDHA's 'Bark of Salvation',

We pray to you that you lead our spirits, safely and without tarry, to beyond the dharma shore of this world's samsaric suffering.

Make haste and make good upon your promise to act as the Sweet Savioress of Mankind, along side Mother Mary and Mother Tara,

And by your Motherly Mercy and Boundless Grace, please help to bring to birth on this ready earth, the Great Golden Age of LORD MAITREYA's Avataric Mission.

Section II

Invocation to Mary
Motherly Matrix of the Divine

O Blessed Mother Mary, Thou in whose womb is carried the child of humanity.

Do bless, O Pure Being of Virgin Blue, this immature child of yet callow heart and (still) selfish aspiration.

May all those who are herewith endowed with the gift of invocative speech,

Invoke thy list of Holy Names now and harken upon Thee for help and mercy.

> O Mary, Mistress of the High Heavens.
>
> O Mary, Mistress of the Tribes of Yore.
>
> O Mary, Mother of Heaven and Earth.
>
> O Mary, Mother of All Who Are Orphans.
>
> O Mary, Mother of Life Abundant.
>
> O Mary, Restorer and Giver of All Life.
>
> O Mary, Blessed Breast in offering to all Babes.
>
> O Mary, Light and Ladder to the Divine.
>
> O Mary, Door to Life Everlasting.
>
> O Mary, All-Seeing, and Merciful Eye of the Sky.
>
> O Mary, Queen of the Isles of the Blessed.
>
> O Mary, Great Womanly Light of the City of Lights.
>
> O Mary, Key to the Kingdom of the Heaven.
>
> O Mary, Protector Behind the Throne of God.
>
> O Mary, Mother of Jesus Mac Mary.
>
> O Mary, Mother Defender of the All Innocents.
>
> O Mary, Metaphorical Madonna of the Ascension.
>
> O Mary, 'Mater Misericordia' Salvatrix.

O Mary, Mother of the Five Joys and (Merciful) Mother of the Five Sorrows.

O Mary, Our Lady of Love and gentle Dakini of Compassion.

O Mary, Handmaiden of Multiform and Manifold Miracles.

O Mary, Mother of Mercy and of Manifest Motherliness.

O Mary, Paragon of Pulchritude and Epitome of Purity.

O Mary, Mother of Untainted Truth and Mother of Heart-Wisdom.

O Mary, Mother of Faith, Hope and Charity.

O Mary, Unconditional Devotee of Devotion (in the LORD's Name).

O Mary, Merciful Intermediary of Heaven (and Hell).

O Mary, 'Le Salut' of Sinners and Redemptress of Reprobates.

O Mary, the Forgiver of Trespasses and Liberator of Transgressors.

O Mary, Crusher of the Serpent and Girder of the Kundalini.

O Mary, Destroyer of the Devil and Upholder of Righteousness.

O Mary, Motherly Defender of the Defenceless and Innocent.

O Mary, Motherly Matrix of Freedom, Liberty and Sovereignty.

O Mary, Blessed Womb of All Women and Promoter of Feminine Emancipation.

O Mary, Queen of Men and Angels Alike and Selfless Servant of Shamballa.

O Mary, Devoted Disciple of the Christ and Maiden Messenger of Maitreya.

And to these laudatory comments ought Humanity to realize in its Collective Mind,

That as long as love, honor and service go out with respect to Maitreya, the Christ,

So too, ought love, honor and esteem blossom forth in the hearts of men for Lady Mary, Mother of the Divine.

Hail to Mary, Mother of the Divine in the illumined hearts of men.

Hail to Mary, Mother of the Divine whispered on wings of angels.

Hail to Mary, Mother of the Divine in the Enlightened Mind of the LORD.

December Full Moon

Little Devi Yogini S.T.R.

A Tibetan 'reincarnation' and a devoted disciple of Djwhal Khul, alias The Tibetan, as well as an Initiate of Sanat Kumara's Ashram.

Little Devi Yogini S.T.R.'s Christmas Wish of Joy for the Whole Planet

Rays 3
2
1

Sanat's (Santa's) Christmas Helper

December Full Moon

I.   A Christmas Celebration

O Silvery Moon of the Christ Child in Joyful Celebration, blast my brain with Thy burst of Christmas Bright.

This Festive Season, please do blitz my mind blind of all hinder and bind.

Blackout all blab and blat, bickering and brawl, and let the Brotherhood of the Jesus Babe, be my only beam.

Bridle my bilge and bigotry and bong bewitchingly kind Thy jingle bells.

Bellow Thy Name brilliantly Lighted and boldly Peaceful into my body beautiful,
And spontaneously burst Thy bedazzling Breath abrim with burning Alleluiahs, into my festive blood.

O Christ Lord, batter my body, break my back, and betroth humbly my 'Xmas bones.

Behold my former blustering bearing as presently self-bankrupt and ego-bare;
Behold my brainsick, desire-blinkings as being now bland and Buddha-blank.

O Prince of Peace, baptize me in this moment as Born-again, deep within the crypt of Thy Benign Bosom.
Bodefully bless my Becoming a better being and bury me in the basic Bliss of Thy bountiful Beneficence.

O Christ Lord, I bid Thee, bedrape me as forever beguiled by the humble beat of Thy loving Heart,
And in the name of Thy Blazing Burnish and Abiding Beauty, let me share in Thy Amazing Grace.

I bid Thee, dear Lord, brazenly bestride me and kind-heartedly, let Thy brassy Beam,
Blush bountifully and graciously broad upon my snow-blazoned, Christmas brow.

II.   **In the December Full Moon of the Festive Season**

> May the Illumined Mind of CHRIST
> light up the minds of all men.
>
> Let us with awareness
> realize our lighted Unity!
>
> GLORY BE TO OUR LORD, THE CHRIST!
> MAY HE BLESS THE WORLD

In the December Full Moon of the Festive Season

May the Loving Heart of CHRIST
Enflame the hearts of all men.

Let us in our heart of hearts,
bond together as One LOVE!

GLORY BE TO OUR LORD, THE CHRIST!
MAY HE BLESS THE WORLD

May the Great Soul of CHRIST
gather together the living souls of all men.

Let us, consciously and collectively,
become the sacred Circle of Humankind!

GLORY BE TO OUR LORD, THE CHRIST!
MAY HE BLESS THE WORLD

May the Evolutionary Spirit of CHRIST
inspire and elevate the spirits of all men.

Let us kindly share the wealth,
participate ecumenically in knowledge,
and cooperatively work the world!

GLORY BE TO OUR LORD, THE CHRIST!
MAY HE BLESS THE WORLD

May the Enlightened Will of CHRIST
fulfill the Purpose and inspirit the Plan
for all men, (and angels alike).

Let us, as a free Mankind
attend to the Purpose of the Father,
and serve the Plan of MAITREYA, the Christ…
for all his children on earth, (as in Heaven).

GLORY BE TO OUR LORD, THE CHRIST!
MAY HE BLESS THE WORLD

In the December Full Moon of the Festive Season (for Buddhists)

May the Illumined Mind
of the BUDDHA light up
the minds of all men.

Let us, in awareness,
realize our lighted Unity!

MAY BUDDHA BLESS THE WORLD!

May the Compassionate
Heart of the BUDDHA
enflame the hearts of all men.

Let us, in our heart of hearts,
bond together as One LOVE.

MAY BUDDHA BLESS THE WORLD!

May the Nonself of the BUDDHA
gather together the selfless selves
of all men.

Let us, collectively and consciously, become the Sacred Sangha of Humankind!

MAY BUDDHA BLESS THE WORLD!

May the Enlightened Spirit of the BUDDHA irradiate and elevate the spirits of all men.

Let us all participate in the five wisdoms, cooperatively work the world,
and kindly share the wealth!

MAY BUDDHA BLESS THE WORLD!

In the December Full Moon of the Festive Season (for Buddhists)

May the Great Will of the Tathagata BUDDHA
inspirit the Purpose and fulfill the Plan
for all men, and devas alike.

Let us as a free Mankind,
attend to the Purpose of the Adi BUDDHA and serve the Plan of MAITREYA,
the Future Buddha… for all his children
on earth, as well as in Tushita Heaven.

MAY BUDDHA BLESS THE WORLD!

III. ## December Full Moon

Let the Pure Mind
at any Time
tote a heart that
mindfully helps.

Let the Kind Heart
bestow a mind
that humanly cares.

Let the Compassionate
Self sport a spirit
full of goodwill.

Let the Shining
Spirit light a candle
of selfless service to all.

Let it be so, at anytime,
especially Now,
under the
Hallowed Radiance
of the Christmas Full Moon.

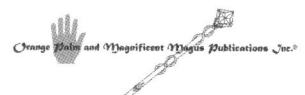

Orange Palm and Magnificent Magus Publications Inc.©
235 René Lévesque Boulevard East, Suite 310
Montréal, Québec, H2X 1N8, Canada
Telephone: (514) 255-8700
Facsimile: (514) 255-0478
E-mail: info@palmpublications.com
Web site: http://www.palmpublications.com